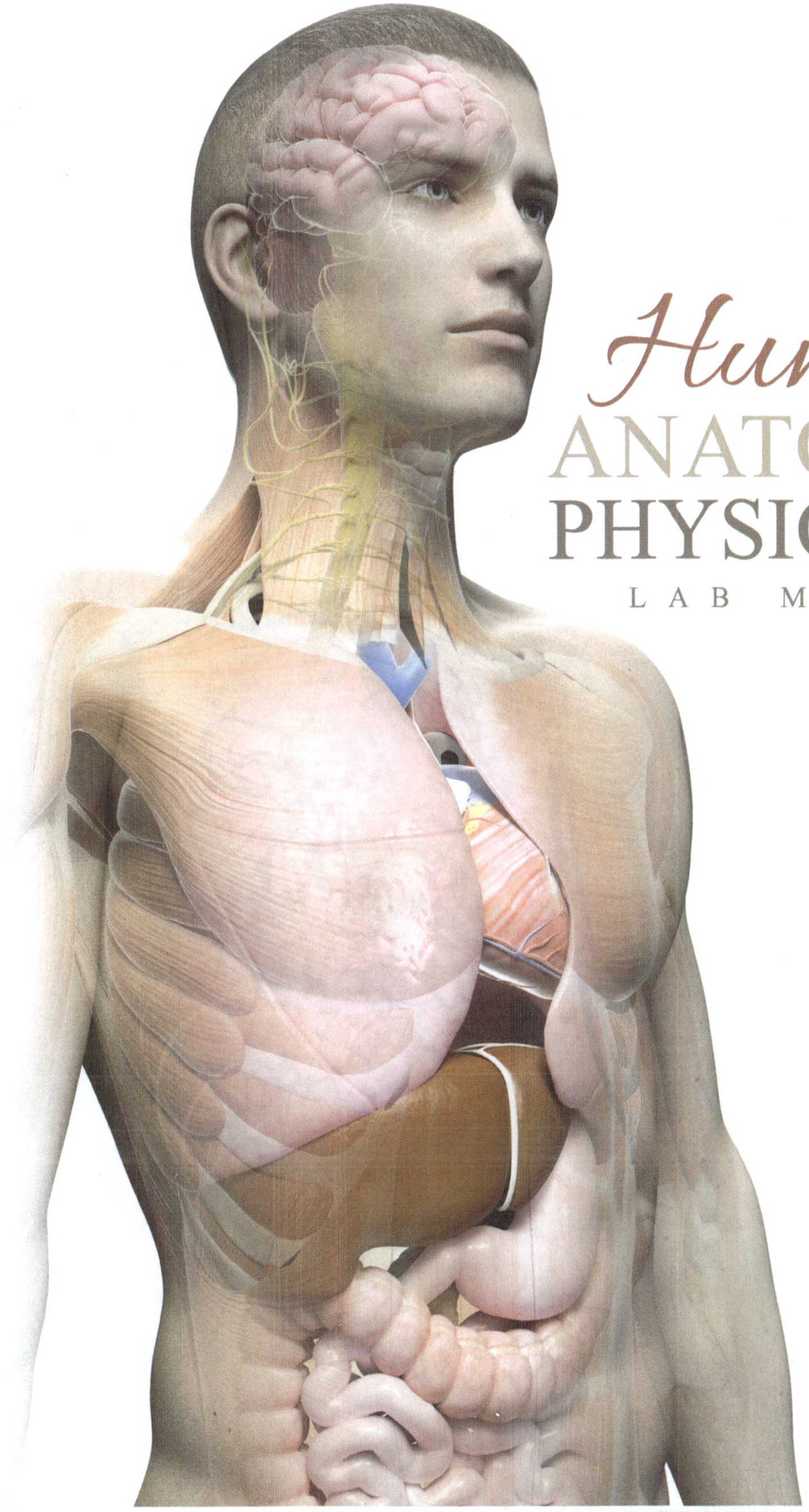

Human ANATOMY & PHYSIOLOGY

LAB MANUAL

JODIE GERTS

Kendall Hunt
publishing company

www.kendallhunt.com
Send all inquiries to:
4050 Westmark Drive
Dubuque, IA 52004-1840

CONTENTS

L A B 1

Introduction, Anatomy Terms, and Introductory Chemistry

Objectives

The purpose of this lab is to introduce the student to the structure and logistics of the lab and to provide terms that serve as the foundation for further study. The student will complete a worksheet that utilizes preliminary concepts and terms that will be used throughout the course and in subsequent health care courses. The student will apply terms related to planes of the body, directional terms and body regions. The student will have an understanding of how anatomical position is used as a reference point. The student will review general chemistry concepts and begin basic biochemistry.

Complete APR Dissection and Animations

During lab: Complete each objective prior to leaving lab.

☐ **Introduction to APR 3.0**

☐ **Worksheet**

Planes

Body Regions

Directional Terms

Synthesis

MODULE 1: BODY ORIENTATION

Dissection

Body Position

LAYER 1: Anatomical Position

LAYER 2: Supine

LAYER 3: Prone

Planes of Section

LAYER 1: Coronal plane, oblique plane, sagittal plane, transverse plane

What is another name for coronal plane? _______________________________

Which plane divides the body into anterior and posterior portions? _______________________________

Which plane is on a slant? _______________________________

Which plane divides the body into left and right portions? _______________________________

Which plane divides the body into superior and inferior portions? _______________________________

Directional Terms

LAYER 1: Anterior, posterior

LAYER 2: Medial, lateral, inferior, superior, distal, proximal

LAYER 3: Deep, superficial

What is synonymous with caudal? _______________________________

What is synonymous with dorsal? _______________________________

What is synonymous with cranial? _______________________________

What is synonymous with ventral? _______________________________

Match the directional terms with their respective description.

_______ Medial	a. Toward the midline
_______ Distal	b. Away from the midline
_______ Inferior	c. Upward
_______ Posterior	d. Downward
_______ Superficial	e. Toward the back
_______ Lateral	f. Toward the front
_______ Superior	g. Closer to the trunk
_______ Anterior	h. Further from the trunk
_______ Proximal	i. Toward the surface
_______ Deep	j. Away from the surface

Body Regions

LAYER 1: Antebrachial, auricular, axillary, brachial, buccal, carpal, cranial, cubital, digits of hand and foot, dorsum of hand and foot, femoral, frontal, gluteal, left and right inguinal, lumbar, nasal, occipital, oral, orbital, palmar, pectoral, perineal, popliteal, sacral, scapular, temporal, thoracic, umbilical, vertebra

Pleura and Pericardium

LAYER 1: Parietal pericardium, parietal pleura, pericardial cavity, pleural cavity, visceral pericardium, visceral pleura

Peritoneum

LAYER 1: Parietal peritoneum, peritoneal cavity, visceral peritoneum

MODULE 2: CELLS AND CHEMISTRY

Animation: Atomic Structure

1. What is the smallest particle of an element that retains the properties of that element? _______________

   ____

2. What comprises the central nucleus? ____

3. What is an electron shell? ____

4. What is the charge of an electron, proton and neutron? ___________________________________

5. What is the atomic number? ____

6. What is the atomic mass? ____

7. What are isotopes? ____

8. How many electrons are held in the second valence shell? ___________________

9. What determines the chemical bonding properties of an atom? ________________

10. Which type of element is most stable and chemically inactive? _______________

11. What determines the number and the kind of bonds an atom forms? ___________

Animation: Bonds

1. Why do atoms gain, share, or lose electrons? ____________________________

2. Which type of bonding shares electrons? _______________________________

3. Which type of bonding donates electrons? ______________________________

4. Why do atoms want to achieve an octet of electrons? _____________________

5. In the example of sodium and chloride bonding, is sodium electrically neutral?   *YES   NO*   (circle)

6. What are the two properties of nonpolar covalent bonds? __________________

7. What is electronegativity? ____

8. What is the difference between nonpolar covalent bonds and polar covalent bonds? __________

   ____

9. In the example of HCl, which atom do electrons spend more time around? ______

10. What does the $\delta+$ symbol mean? ___________________________________

Animation: Electrolytes

1. Define electrolyte. ____

2. What makes sodium chloride a strong electrolyte? _______________________

3. What does the positive end of the water interact with? ____________________

4. Define dissociate. ____

5. Give an example of a weak acid. What makes it a weak electrolyte? ___________

   ____

6. What is the formula for carbonic acid? _________________________________

7. What is the formula for bicarbonate? ____

8. Does glucose dissolve into ions in water? Is glucose an electrolyte? Can glucose dissociate? _______________

____

9. The hydroxyl group [-OH] of glucose is attracted to the _______________ of other glucose

   molecules and the _______________ of water. This attraction is called a _______________

   bond. Which is stronger, the attraction to other glucose molecules or the attraction to water

   molecules? *Glucose   Water*   (circle)

10. Can glucose dissolve in water? *YES   NO*   (circle)

   Can glucose dissociate in water? *YES   NO*   (circle)

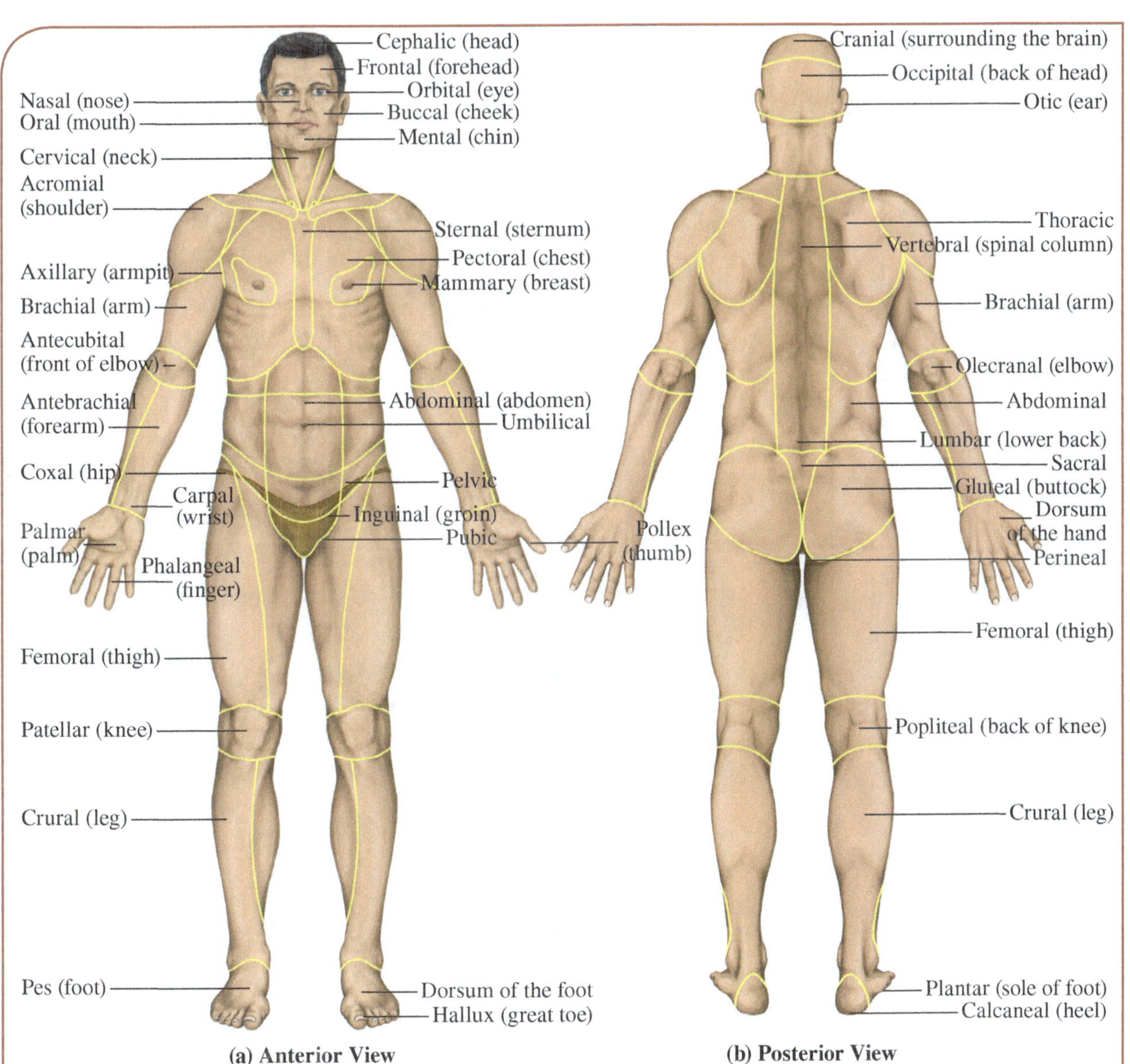

© stihii/Shutterstock.com

BODY REGIONS

Select from the following body regional terms to be used in Part 2 of Worksheet (*Not all will be used*)

Abdominal	Frontal	Pectoral
Acromial	Gluteal	Pelvic
Antebrachial	Inguinal	Perineal
Antecubital	Hallux	Phalangeal (hand and foot)
Axillary	Lumbar	Plantar
Brachium/Brachial	Mammary	Pollex
Buccal	Mental	Popliteal
Calcaneal	Nasal	Pubic
Carpal	Occipital	Sternal
Cervical	Olecranal	Sacral
Coxal	Oral	Tarsal
Cranial	Orbital	Thoracic
Crural	Otic	Umbilical
Dorsum (hand and foot)	Palmar	Vertebral
Femoral	Patellar	

LAB 1 ASSIGNMENT

Turn in at the end of lab.

Name: ____   Lab Day / Time: _____________________

PART 1: Planes

(circle the letter)

1. A   B   C   D   4. A   B   C   D   6. A   B   C   D
2. A   B   C   D   5. A   B   C   D   7. A   B   C   D
3. A   B   C   D

PART 2: Body Regions

Select from the terms listed in your Lab Manual

1. _______________________   16. _______________________
2. _______________________   17. _______________________
3. _______________________   18. _______________________
4. _______________________   19. _______________________
5. _______________________   20. _______________________
6. _______________________   21. _______________________
7. _______________________   22. _______________________
8. _______________________   23. _______________________
9. _______________________   24. _______________________
10. _______________________   25. _______________________
11. _______________________   26. _______________________
12. _______________________   27. _______________________
13. _______________________   28. _______________________
14. _______________________   29. _______________________
15. _______________________   30. _______________________

PART 3: Directional Terms

List the correct answer from the following terms based on the list.

Medial	Proximal	Superior
Anterior	Lateral	Distal
Superficial	Posterior	
Inferior	Deep	

1. B is ___________ to A.

2. D is ___________ to E.

3. F is ___________ to G.

4. C is ___________ to B.

5. H is ___________ to I.

6. J is ___________ to K.

7. L is ___________ to M.

8. L is ___________ to N.

9. M is ___________ to N.

10. P is ___________ to O.

PART 4: Synthesis

In the following pictures, use the descriptions to determine the location of each traumatic event for your patients who just came into your emergency department.

PATIENT #1: Designate with the following symbols where the entrance (X) and exit (+) wound is located.

Entrance wound: Patient has an entrance wound from a gunshot in the right pectoral region 1 inch medial to the axillary region. Mark with an (X).

Exit wound: The bullet exited the patient in the left posterior thoracic region 1 inch inferior to the scapular region and 2 inches lateral to the vertebral region. Mark with a (+).

PATIENT #2: Shade in the areas where the burn is located. Patient has an extensive burn from an explosion in the following regions: left posterior crural, left popliteal, left posterior femoral, dorsum of left hand, left pollex, left posterior brachium, left olecranal, left lateral brachium, left lateral cervical, left buccal, left otic regions.

LAB 2

Biochemical Molecules, Cellular Structures, pH, and Microscopy

Objectives

The purpose of this lab is to have the student learn how to use the microscope and identify biochemical molecules and cellular structures. The student will also perform experiments to help them understand acids and bases and quantification using the pH scale.

Prior to lab: Complete the Pre-lab.

Complete APR Dissection, Histology, and Animations

During lab: Complete each objective prior to leaving lab.

- [ ] **Cell Model**
     Identify Cell Structures and Their Functions
- [ ] **Identification of Biochemical Molecules**
- [ ] **Building a Sucrose Model**
- [ ] **Human Cheek Cell Microscope Experiment**
     Introduction to the Microscope
- [ ] **pH Experiment**
     Sample Liquid Experiment
     Saliva
     Stomach pH

LAB 2 PRELAB—Biochemical Molecules, Cellular Structures, pH, and Microscopy

PART 1:

Name the element represented by each chemical symbol and state the number of valence electrons using a periodic table.

Chemical Symbol	Element	Atomic Number	Number of Valence Electrons
Na			
Fe			
Ca			
K			
Cl			
C			
P			
N			
H			
O			
S			
Mg			

PART 2:

Label each functional group.

Symbol	Name
-OH	
$-NH_2$	
-COOH	
$-CH_3$	
$-H_2PO_4$	

PART 3:

Which of the following are monosaccharides, disaccharides, polysaccharides, or enzymes.

Organic Molecule	Type
Glucose	
Lactase	
Maltose	
Cellulose	
Sucrose	
Galactose	
Sucrase	
Fructose	
Glycogen	
Lactose	
Maltase	

PART 4:

Draw the generic structure of an amino acid using a central carbon, hydrogen, amino group, carboxyl group, and R group.

1. Are the R groups nonpolar, polar, or it depends on the type of amino acid? _______________________

2. How many amino acids are there? _______________ How many of those are essential?

   _________________ nonessential? _________________

PART 5:

Dra Identify the following biomolecular structural formulas from this list of biomolecules.

Carbohydrates	Lipids	Proteins	Nucleotides/Nucleic Acids
Monosaccharides	Phospholipid	**Amino acids**	**Nitrogenous bases**
Fructose	Cholesterol	Alanine	Guanine
Glucose	Glycerol	Glycine	Cytosine
Disaccharides	Fatty acid	Cysteine	**Nucleotides/Nucleic Acids**
Maltose	Saturated triglyceride	Dipeptide (ala-gly)	ATP
Sucrose	Unsaturated triglyceride	Protein	RNA
Polysaccharides			DNA
Glycogen			

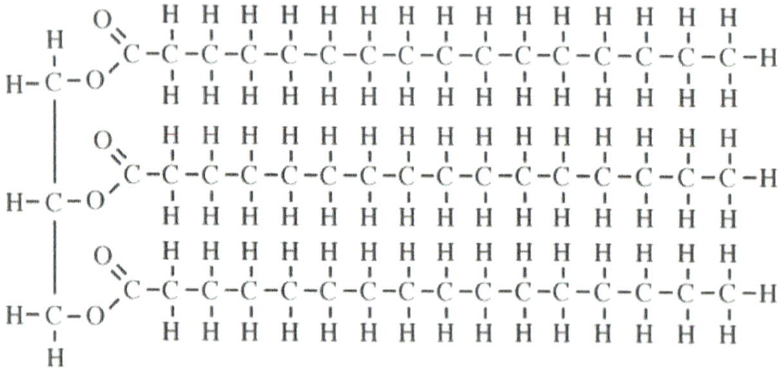

1. ____

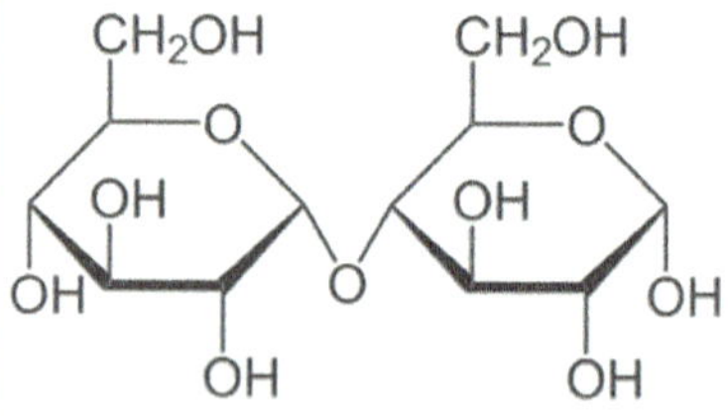

2. ____

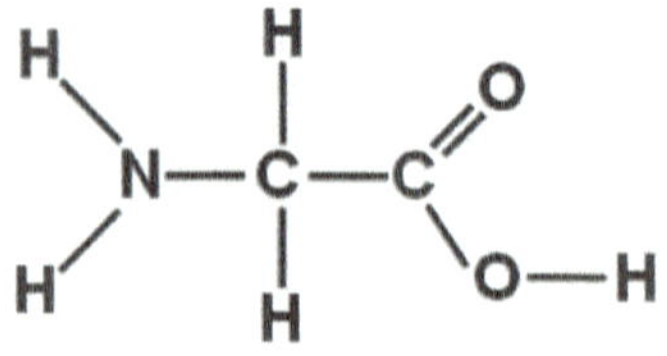

3. ____

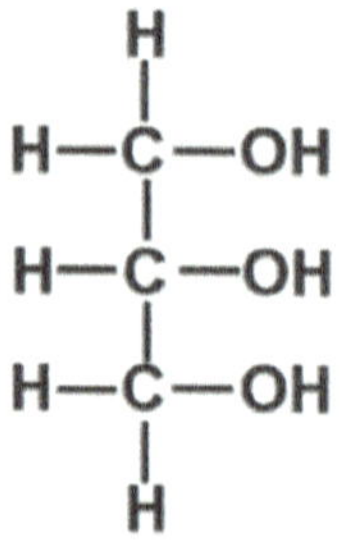

4. ____

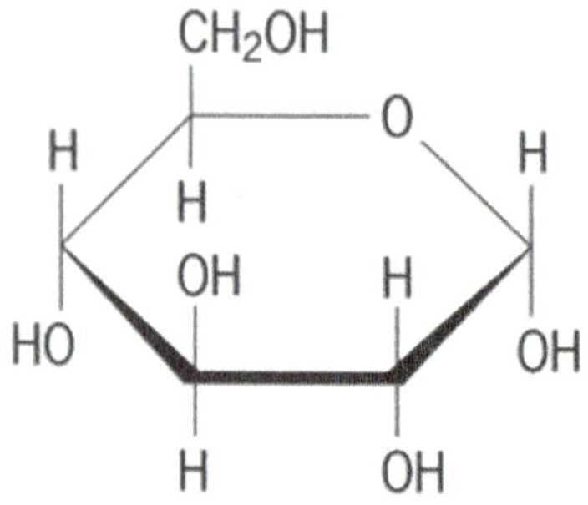

5. ____

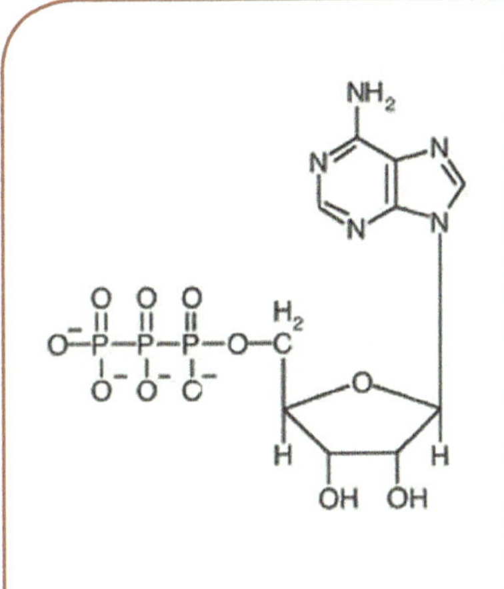

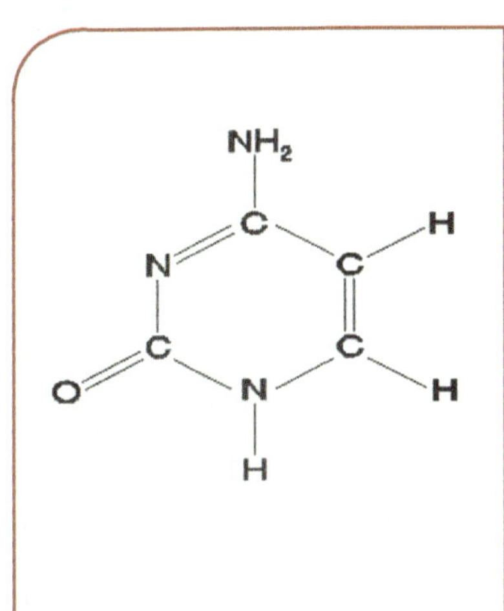

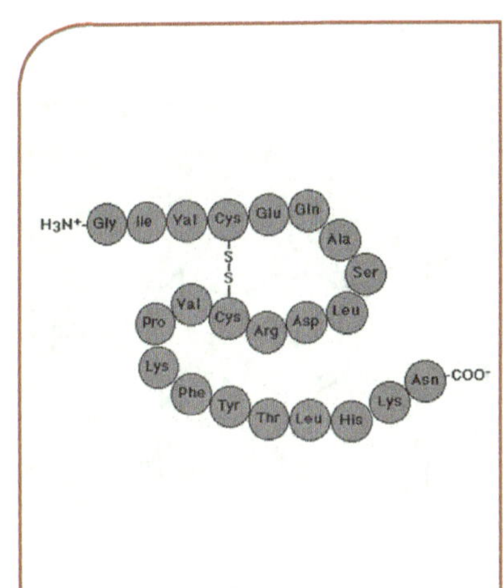

6. _______________________

7. _______________________

8. _______________________

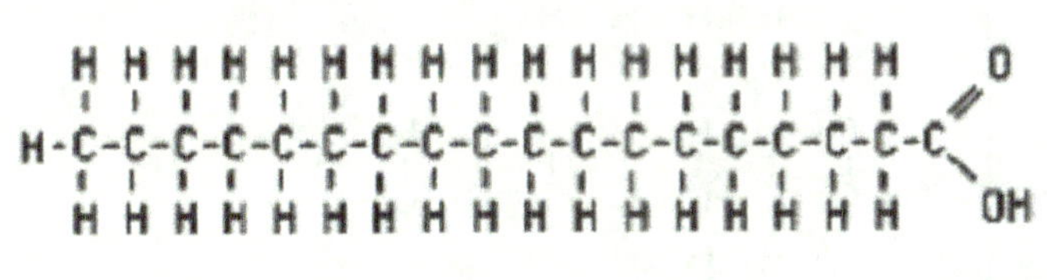

9. _______________________

10. _______________________

11. _______________________

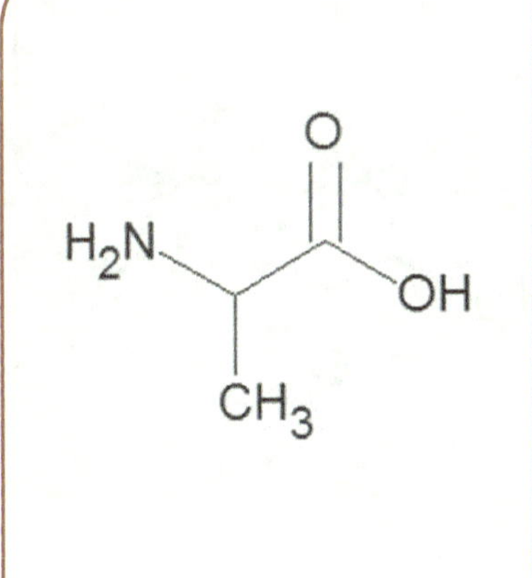

12. _______________________

13. ________________________________

14. ________________________________

15. ________________________________

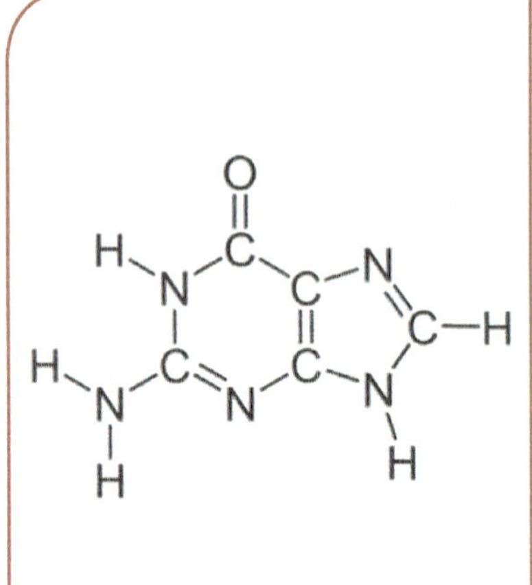

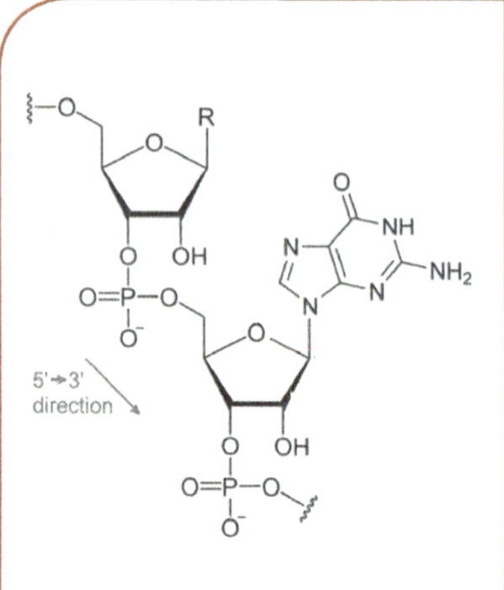

16. ________________________________

17. ________________________________

18. ________________________________

19. ________________________________

20. ________________________________

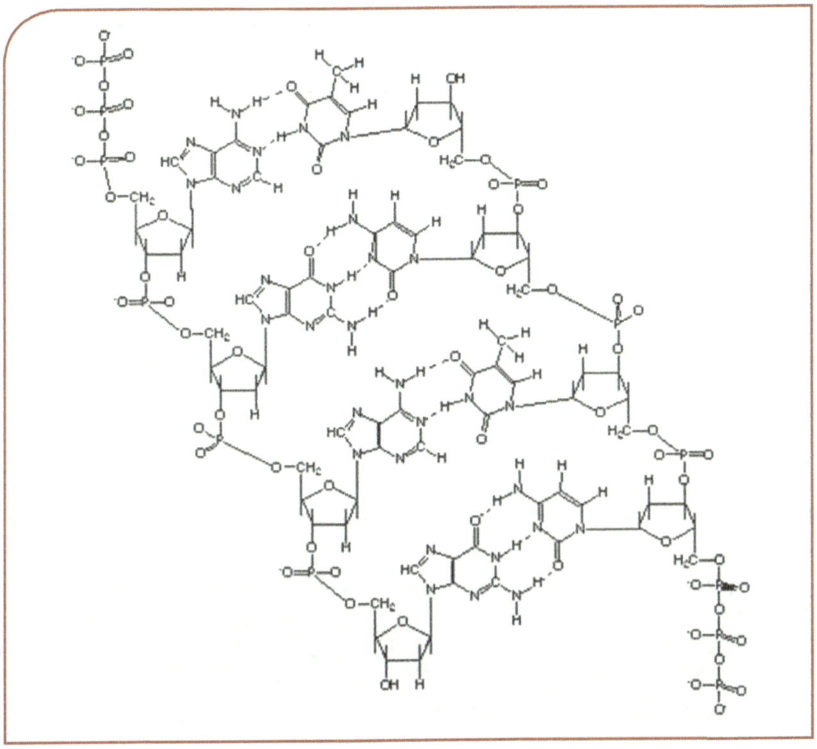

21. ____

PART 6:

What stands out that makes each of these biomolecules readily identifiable?

CARBOHYDRATES: To help you identify carbohydrates, find common characteristics among carbohydrates and unique characteristics that help you to differentiate the molecules from each other and other biomolecules.

1. What basic elements are found in all these carbohydrate examples? _______________________________

2. Fructose and glucose are very similar but what is the difference? _______________________________

3. Which molecule or elements are missing from maltose versus two independent glucose molecules?

4. Glycogen is a polymer of _______________ molecules.

LIPIDS: To help you identify lipids, find common characteristics among lipids and unique characteristics that help you to differentiate the molecules from each other and other biomolecules.

1. What basic elements are found in all these lipid examples? _______________________________

2. Which molecule has a four ring structure? _______________________________

3. Which lipid molecule has a phosphate group? ____

4. Which molecules have a glycerol? ____

5. Which molecules have fatty acids? ____

6. What is the difference between a saturated and an unsaturated fatty acid? ____________________

____

PROTEINS: Find common characteristics among proteins and unique characteristics that help you differentiate the molecules from each other and other biomolecules.

All amino acids have a minimum of four elements…what are they?

_______________________________   _________________________________

_______________________________   _________________________________

7. What distinguishes one amino acid from another? ____

NUCLEOTIDES/NUCLEIC ACIDS: Find common and unique characteristics that can help you to identify these molecules.

1. Name three principle components of nucleotides.

____

____

____

2. What type of molecule is adenine (from adenosine triphosphate)? ___________________________

3. How can you identify an ATP from a phospholipid? _______________________________________

4. What is the difference between purine and pyrimidine? _____________________________________

5. How can you identify a nitrogenous base (i.e. cytosine) from an amino acid? __________________

6. What type of molecule is ribose and deoxyribose? ____

7. What type of bond is created at the nitrogenous bases between the complementary base pairs?___________

____

8. List the complementary base pairing of the nitrogenous bases.

Uracil: _______________________________________

Guanine: _______________________________________

Adenine: _______________________________________

Thymine: _______________________________________

Cytosine: _______________________________________

PART 7:

Identify these structures on the cell.

Centrioles

Chromatin

Endocytosis

Exocytosis

Free ribosomes

Golgi apparatus (Golgi complex)

Lysosomes

Mitochondria

Nuclear pores

Nucleus

Peroxisomes

Plasma membrane

Rough endoplasmic reticulum

Secretory vesicles

Smooth endoplasmic reticulum

PART 8:

Answer the following questions about the cell.

1. Which organelle is responsible for ATP production? _______________________________

2. Which organelle in the cell produces lipids? _______________________________

3. Which organelle in the cell breaks down long fatty acids? _______________________________

4. The DNA that controls the functions of the cell is located in which cellular structure? _______________

5. Which structure in the cell stores calcium? _______________________________

6. Which type of ribosome (free or rough) produces proteins used exclusively inside the cell? _______________

    ____

7. Which organelle breaks down bacteria, virus, and other organelles? _______________________________

8. What is the function of the cytoskeleton? _______________________________

9. Cholesterol is vital for the structural integrity of which cellular structure? _______________________________

10. Which structure in the cell is mostly composed of a phospholipid bilayer? _______________________________

11. The liquid portion of the cytoplasm is called ____.

12. Through what structure does mRNA exit the nucleus? ________________________________

13. What is the compact form of chromatin called? ____________________________________

PART 9:

Answer the following questions about pH.

1. What is the range of the pH scale? ____

2. What pH is distilled water? ____

3. Distilled water is a neutral pH which means that there is as many negative _________________ ions as

    there are positive _________________ ions.

4. Another name for alkaline is ____.

5. What pH range is used for acidic solutions? ____

6. The higher the pH, the more _________________ the solution.

7. What is the normal pH of our blood plasma? _________________ Is this acidic or alkaline?

    ____

8. Write out the most common reversible reaction in the body. ____________________________

    ____

9. In reference to the above equation, when we say that the equation goes "to the right" it means that there is

    an abundance of _________________ molecules which drive the reaction. When the equation "goes to

    the left" it means that there is an abundance of _________________ ions driving the reaction. Reversible

    reaction follows the Law of _________________. Net reactions occur until equilibrium is reached.

10. When someone has excessive CO_2 in their blood, this would drive the above reaction to

    the   *right*   *left*   and cause an   *acidic*   *alkaline*   condition in the blood.   (circle)

11. Define dental plaque. What gets rid of dental plaque? _________________________________

    ____

12. What is tartar? What gets rid of tartar? ____

    ____

MODULE 2: CELLS AND CHEMISTRY

Dissection

Generalized Cell

LAYER 1: Cilium, microvilli

LAYER 2: Plasma membrane, microtubule, mitochondrion, centrosome, free ribosome, lysosome

LAYER 3: Smooth endoplasmic reticulum, rough endoplasmic reticulum, golgi apparatus, nucleus, nuclear pore

Plasma Membrane

LAYER 1: Glycocalyx

LAYER 2: Peripheral membrane protein, integral membrane protein, cholesterol molecules, lipid bilayer, polar heads of lipid bilayer, fatty acid tails of lipid bilayer

Histology: Plasma Membrane

Cytoplasm, extracellular matrix, plasma membrane

What is the glycocalyx? ____

Histology: Cytoplasm

Cytoplasm, nucleus

Histology: Rough Endoplasmic Reticulum (High Magnification)

Membrane bound ribosome, free ribosome, rough endoplasmic reticulum

Histology: Smooth Endoplasmic Reticulum

Smooth endoplasmic reticulum

Histology: Golgi Apparatus

Golgi apparatus, secretory vesicle, transport vesicle

What transports proteins from RER to Golgi apparatus? ______________________________

What carries protein products from Golgi apparatus to cell surface? ______________________________

Histology: Mitochondrion

Cristae of mitochondrion, mitochondrial matrix, mitochondrion

What are the cristae? ____

Histology: Cytoskeleton

Intermediate filament, microfilament, microtubule

What is the microfilament also known as? ____

Which cytoskeletal filament makes up the core of the microvilli? ____

Which cytoskeletal filament is located in the centrosome? ____

Histology: Flagellum of Sperm

Flagellum of sperm

Histology: Microvillus (Longitudinal Section)

Microfilament, microvillus

Histology: Cilium and Microvillus

Cilium, microvilli

Which is motile?   *Cilia   Microvilli*   (circle)

Histology: Nucleus

Nuclear envelope, nucleus

Histology: Chromosome

Centromere, chromatid, Metaphase chromosome

How many chromosome pairs are there in the human somatic cells? ____

Animation: Enzymes

1. What is the special region on the enzyme that binds with specific substrate molecules? ____

2. What are the names of the molecules that bind to the enzyme? ____

3. What happens when the enzyme stresses the chemical bonds in the substrates?_______________

4. What happens to the enzyme when the product is released from the active site? _______________

Animation: Cell Cycle and Mitosis

1. What events are included in the cell cycle? _______________

2. Define cytokinesis. _______________

3. What phase of the cell is involved with growth and metabolic activities? _______________

4. In which phase of interphase do normal cellular activities occur? _______________

5. What does the S phase do? _______________

6. What are the four phases of mitosis?

_______________ _______________

_______________ _______________

7. What happens in prophase? _______________

8. What is each identical copy of a single chromosome called? _______________

9. What grows out of the centrioles to form the spindle fibers? _______________

10. At which phase does the chromosomes line up at the center of the cell? _______________

11. What attaches to the sister chromatids in metaphase? _______________

12. When the sister chromatids pull apart, it is called _______________.

13. What divides in the double-stranded chromosome during anaphase? _______________

14. In which phase of mitosis do the chromatids uncoil? _______________

15. Which events result in two genetically identical daughter cells? _______________

Animation: DNA Structure

1. DNA has _______________ strands of nucleotides.

2. A nucleotide is composed of three things. What are they?

_______________ _______________ _______________

3. What forms the backbone or sides of the ladder of DNA? _______________

4. What occurs between the complementary nitrogenous bases of each strand of DNA? _______________

____

5. What are the four nitrogenous bases?

______________________________________   ______________________________________

______________________________________   ______________________________________

Animation: Lysosomes

1. What types of enzymes are found in lysosomes?____

2. What do these enzymes break down? ____

3. Where are the enzymes that are found in lysosomes made? _________________________________

4. Transport vesicles go from the ___________________ to the ___________________.

5. Lysosomes pinch off from the ____.

6. When a cell ingests viruses or bacteria, what fuses with that vesicle? _________________________

7. Within the cell, what destroys the virus or bacterium? ____________________________________

Animation: DNA Replication

1. What is DNA replication? ____

2. What is the name of the original strands of DNA? ____

3. The bonding of adenine to thymine or cytosine to guanine is called _________________________.

4. The two new DNA molecules come from an ___________________ strand and a

___________________ strand.

Animation: Diffusion

1. The eventual result of constant kinetic energy of dissolved moles in a solution is ________________.

2. Define diffusion. ____

3. Are molecules in motion in a solid?   *YES   NO*   (circle)

4. What happens when a sugar cube is placed in water?____

5. Do the molecules move from a region of high concentration to low concentration or from an area of low concentration to high concentration? ____

6. When does diffusion stop?____

7. Which three things affect the rate of diffusion?

___________________________   ___________________________   ___________________________

Animation: Osmosis and Tonicity

1. Define osmosis. ____

2. Define tonicity. ____

3. What does "relative concentrations" mean? ____

4. Tonicity describes the   *solution   cell*   . (circle)

5. Give an example of a hypertonic solution relative to the normal cell? ____

6. What is normal saline? ____

7. Does water move across the plasma membrane in an isotonic solution? Is there a net movement of water?

____

8. A hypertonic solution causes the cell to _________________ and thus the cell _________________.

9. Define crenation. ____

10. In a hypotonic solution, the solute concentration is   *greater   less*   inside the cell than in the extracellar fluid.   (circle)

11. A hypotonic solution causes a net movement of water _________________ the cell and the result is that the cell   *crenates   swells*   . (circle)

12. Define lyse. ____

13. What is the difference between lysis and hemolysis? ____

MODELS: IDENTIFICATION OF BIOCHEMICAL MOLECULES

LIPIDS	CARBOHYDRATES	PROTEINS	NUCLEOTIDES/ NUCLEIC ACIDS
1. Triglyceride	6. Glucose	10. Alanine (amino acid)	14. ATP
2. Phospholipid	7. Fructose	11. Glycine (amino acid)	15. Purine
3. Cholesterol	8. Maltose (glu+glu)	12. Cysteine (amino acid)	16. Pyrimidine
4. Glycerol	9. Sucrose (glu+fru)	13. Dipeptide (glycine + alanine)	17. DNA
5. Fatty acid			

BIOCHEMICAL IDENTIFICATION

Organized into four stations are different organic molecules that are common in the human body. We will be discussing these molecules for the rest of the year. It is essential you recognize their structure and to help understand their properties. The stations are divided into lipids, carbohydrates, proteins, nucleotides/nucleic acids. I have given you a chemical structure and you must identify the molecule based on the structure. Use the Color Coded Key to identify the atoms.

Color Key	
Carbon	Black
Oxygen	Red
Hydrogen	White
Nitrogen	Blue
Sulfur	Yellow
Phosphorus	Purple

Name these molecules from the list. This list is not all inclusive of the molecules displayed.

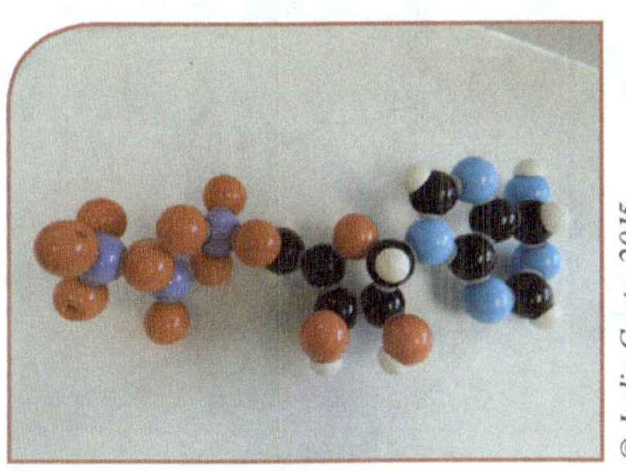

© Jodie Gerts, 2015

© Jodie Gerts, 2015

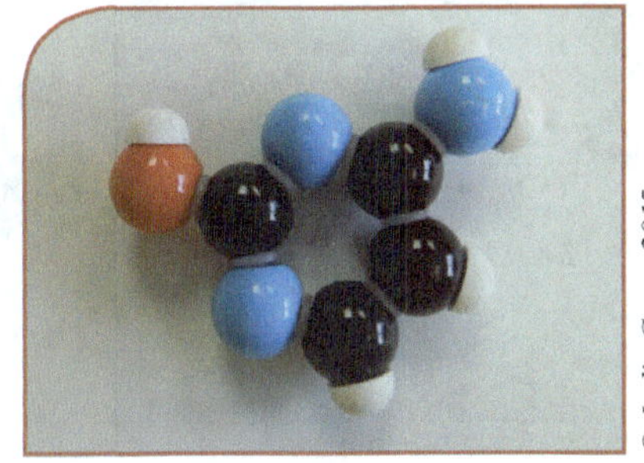

© Jodie Gerts, 2015

1. ___________________   2. ___________________   3. ___________________

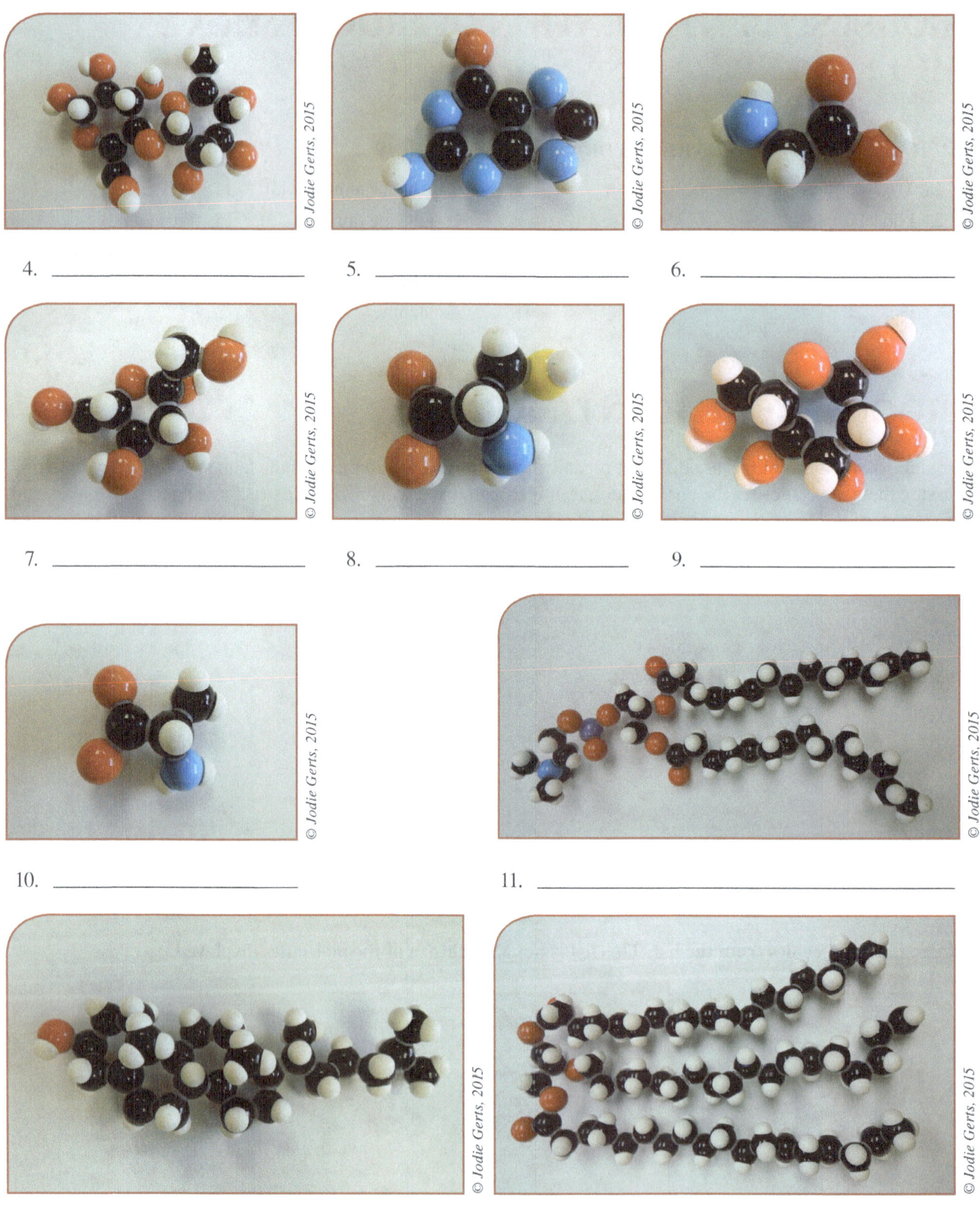

4. _______________________   5. _______________________   6. _______________________

7. _______________________   8. _______________________   9. _______________________

10. _______________________   11. _______________________

12. _______________________   13. _______________________

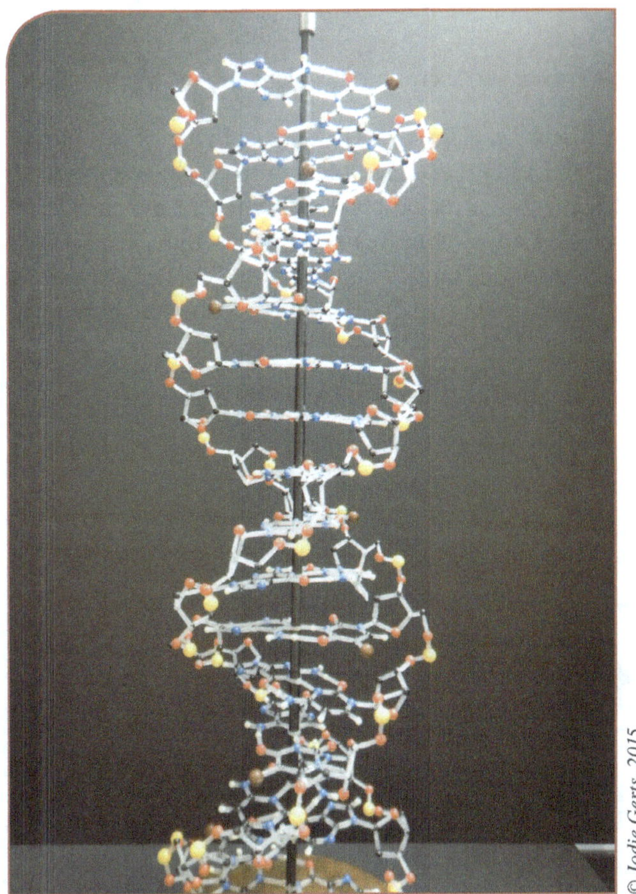

14. _______________________________

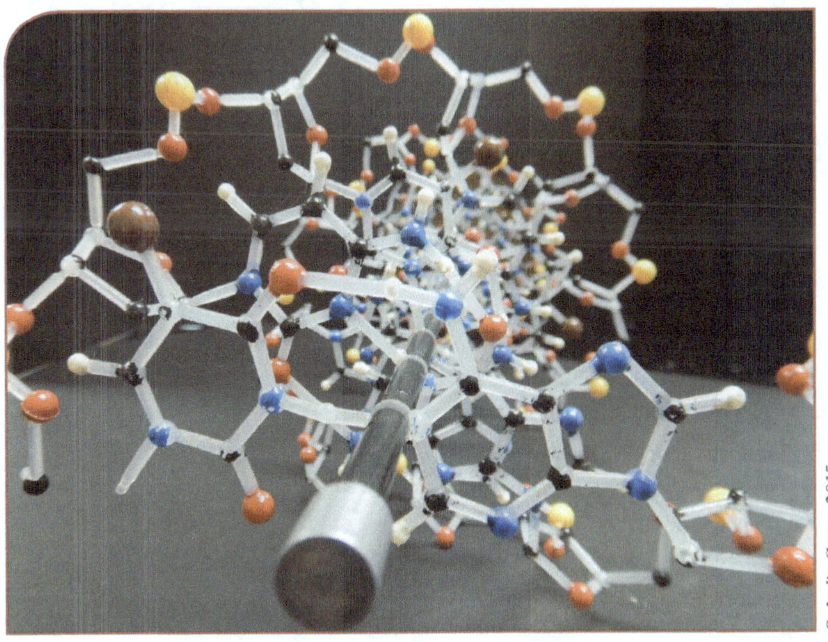

Identify the nitrogenous bases (purine, pyrimidine) and sugar-phosphate backbone (deoxyribose and phosphate group). Also locate the hydrogen bonds between the purine and pyrimadine.

15. _______________________________

Building a Sucrose Model

Today we will build a fructose, glucose, and sucrose molecule using the kit provided.

The sugar we use to sweeten our foods (i.e. table sugar) is sucrose. In the United States, it is estimated that each person consumes an estimated weight of 150 lbs of sugar. Most of the sugar is embedded in foods. Sucrose is a disaccharide molecule composed of a glucose molecule bonded to a fructose molecule. The glucose molecule is a six atom ring structure and the fructose is a five atom ring structure. Both have a formula of $C_6H_{12}O_6$ and are isomers of each other. When glucose and fructose link together, they undergo a dehydration reaction and form the products of sucrose and water. After sucrose enters the body, it is hydrolyzed to glucose and fructose by the enzyme sucrase.

Contents in the Molecular Model Kit		
Quantity	**Atom**	**Color**
12	Carbon	Black
12	Oxygen	Red
24	Hydrogen	White
Short Link Removal Tool		

Building the Glucose Molecule

1. Using the structural formula as a guide begin by building the ring structure of glucose using five carbons and one oxygen to form a six atom ring structure.

2. Next add the additional carbon to the glucose ring and then add the oxygen molecules. Do not be concerned as to the orientation of the oxygen molecules.

3. Add the hydrogen molecules to complete the molecule.

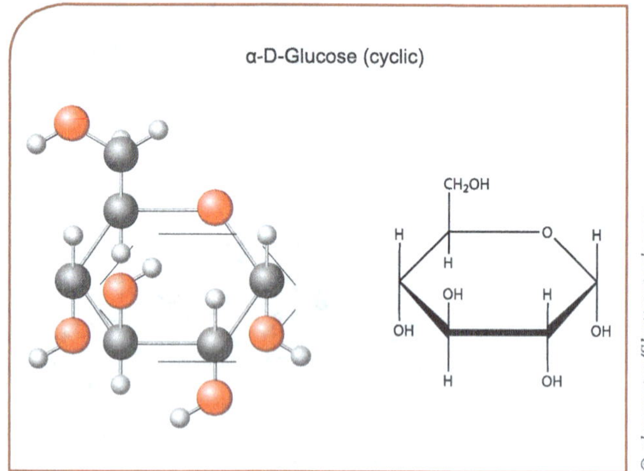

© chromatos/Shutterstock.com

Building the Fructose Molecule

1. Use the structural formula as a guide, begin by building the ring structure of fructose using four carbons and one oxygen to make a five atom ring.

2. Next add the additional two carbons to the fructose molecule and then add the oxygen molecules. Do not be concerned as to the orientation of the oxygen molecules.

3. Add the hydrogen molecules to complete the molecule.

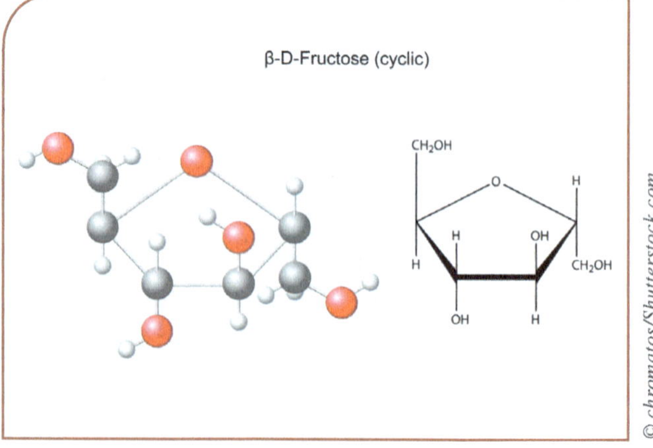

© chromatos/Shutterstock.com

Building the Sucrose Model

1. The glucose and the fructose form a link due to a dehydration reaction. Extract a hydroxyl group (-OH) off the first carbon on the glucose molecule

2. Extract a hydrogen atom off the oxygen of the fructose molecule.

3. Bond the glucose and the fructose to make a sucrose molecule.

4. Bond the hydrogen to the hydroxyl to make a water molecule.

Disassemble the molecule into its individual parts. Use the short link removal tool to assist you. Carefully organize the parts into the box so the next student can use the kit.

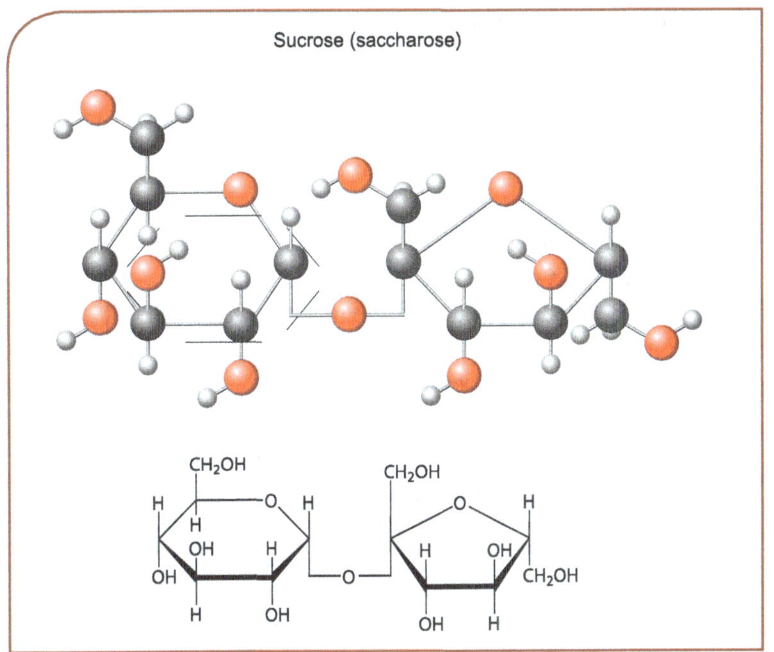

© chromatos/Shutterstock.com

THE HUMAN CHEEK CELL MICROSCOPE EXPERIMENT (WORK INDIVIDUALLY)

In this experiment, we will visualize a type of cell found on the inside of our cheek. These specific cells are called epithelial cells and are easy to obtain by scraping the inside of your cheek with a toothpick. We will stain the cell with a dye called methylene blue which highlights components of a cell. Once the cell is stained, you will utilize a microscope to observe individual cells. A primary purpose of this experiment is to learn to use the microscope.

Materials

Compound light microscope	Cover slip
Methylene blue	Kim-wipe
Toothpick	Water with dropper
Glass microscope slide	

Procedure

1. Place one drop of water on the slide.

2. Scrape the inside of your cheek with the flat side of a toothpick.

3. Mix the toothpick in the water to remove the cells from the toothpick. Throw the toothpick in the biohazard bag.

4. Place a cover slip on the wet part of the slide.

5. Place a drop of Methylene blue on the side of the cover slip. Caution: Methylene blue can stain clothes and skin.

6. Place a Kim-wipe on the opposite side of the drop to draw the Methylene blue under the coverslip. This will soak up the water and draw the Methylene blue under the cover slip. This is called wicking.

7. Place your slide on the stage of the compound microscope and secure it with the lever.

8. Look through the microscope with the 4x objective and focus (lowest power). Cells should be visible, but they will be small and will look like nearly clear purplish pinpoints. If you are looking at large, very dark purple spots, it is probably not a cell.

9. Switch to 40x high power and refocus. Draw the cell. **Label the nucleus, cytoplasm, and cell membrane of a single cell.** Draw your cells to scale.

High Power

pH EXPERIMENT

The pH is a measure of the concentration of hydrogen ion present in a solution. As the hydrogen ion concentration increases, the solution becomes more acidic. As the hydrogen ion concentration decreases, the solution becomes more alkaline or basic. The pH scale ranges from 0 (most acidic) to 14 (most basic). A pH of 7 is considered to be neutral. The pH isn't the actual hydrogen ion concentration; instead it is a negative logarithm of a solution's hydrogen ion concentration. Acids and bases are some of the most important chemicals in the body. They are found in the stomach, blood, urine, cytosol, and other fluids.

- The normal pH for blood is about 7.35–7.45.
- The body regulates this tightly. Even swings of 0.2 points in either direction can have serious consequences.

Sample Liquid Experiment (WORK IN PAIRS)

A simple way to measure pH is to use pH paper. Today you will be testing several solutions. Dip the pH paper into the samples. Saturate the pH strip and compare it to the corresponding colors on the pH box.

Sample Liquid Experiment		
Samples	pH Reading	Is it an Acid or a Base?
Milk (on table)		
Juice (on table)		
Listerine or mouthwash (on table)		
Your drink		

Enamel on your teeth gets eroded with food/drink below a pH level of 5.5. To make matter worse, these acids from food and drink are present in your mouth for about 30 minutes after eating.

Saliva Experiment (WORK INDEPENDENTLY)

1. Spit into a Dixie cup. Take your saliva pH and record.

    pH of saliva: _____________________ **pH of drink:** _____________________

    WHAT DO YOU EXPECT YOUR pH of saliva to be after you drink your acid beverage?

    pH of saliva   AVERAGE pH of saliva + drink   pH of the drink   (circle)

2. Take a gulp of your drink and swish it around in your mouth and then swallow it.
3. Wait 30 seconds to 1 minute.
4. Spit in spit cup again.
5. Measure the pH.

What was the pH of your saliva after swishing with acid? _____________________

What is the difference between your saliva prior to drinking and after drinking? _____________________

Explain your results.

____

____

____

____

____

____

____

____

____

____

____

____

____

STOMACH pH (WORK IN PAIRS)

Now let's apply pH to a pharmacologic example. The parietal cells of the stomach make hydrochloric acid (HCl). HCl is used to denature proteins and activate enzymes in the stomach. Antacids are a group of medicines that neutralize stomach acids to treat a variety of conditions, namely gastroesophageal reflux. We will test three widely known medications for treating reflux: Tums, Alka-Seltzer and Prilosec.

Which drug would you recommend to a patient based on your experiment? ________________________

1. Obtain three 5 mL beakers with 0.1 HCl (stomach acid).
2. Test the pH of one of the beakers. What is the pH? ________________________
3. Add a pinch of crushed Tums to Beaker 1. Stir using a toothpick.
4. Add a pinch crushed Alka-Seltzer to Beaker 2. Stir using a toothpick.
5. Add ½ of a tablet of Prilosec to Beaker 3. Stir using a toothpick.
6. Allow tubes to sit for 1–2 minutes.
7. Measure the pH of the contents of each tube and record the values.

Samples	Active Ingredient	pH
Hydrochloric acid 0.1 M	——————	
HCl and Tums	Calcium carbonate	
HCl and Alka-Seltzer	Sodium bicarbonate	
HCl and Prilosec	Omeprazole	

What drug will you prescribe for your patient suffering with heartburn? ________________________

What drug will you prescribe for your patient suffering with a stomach ulcer? ________________________

L A B 3

Tissues and Protein Synthesis

Objectives

The purpose of this lab is to have the student recognize different tissue types and their associated functions. The student will be able to transcribe and translate DNA to make a protein.

Prior to lab: Complete the Pre-lab.

Complete APR Histology and Animations

During lab: Complete each objective prior to leaving lab.

☐ **Microscopy of Tissues**
    Stratified Squamous Epithelium
    Pseudostratified Epithelium
    Areolar Connective Tissue
    Dense Regular Connective Tissue
    Hyaline Cartilage
    Elastic Cartilage
    Fibrocartilage
    Blood
    Compact Bone
    Skeletal Muscle
    Smooth Muscle

☐ **Identify Cells in Your Blood Sample**

☐ **Protein Synthesis Worksheet**

LAB 3 PRELAB—Tissues and Protein Synthesis

1. List the four types of tissue.

   _____________________________   _____________________________

   _____________________________   _____________________________

2. Label the types of epithelial tissue from the pictures.

   A. ____

   B. ____

   C. ____

   D. ____

   E. ____

3. State which epithelial tissue performs the function listed.

   A. Forms the alveoli of the lungs to permit diffusion of gases: _____________________

   B. Forms the epidermis of the skin: _____________________________________

   C. Permits stretching of the urinary bladder: _______________________________

   D. Secretes hormones from the thyroid gland: _______________________________

4. List the types of connective tissue.

   Connective tissue proper

   1. ____

   2. ____

Supporting connective tissue

   1. ____

   2. ____

Fluid connective tissue

   1. ____

5. There are three types of cartilage tissue. Label the following pictures:

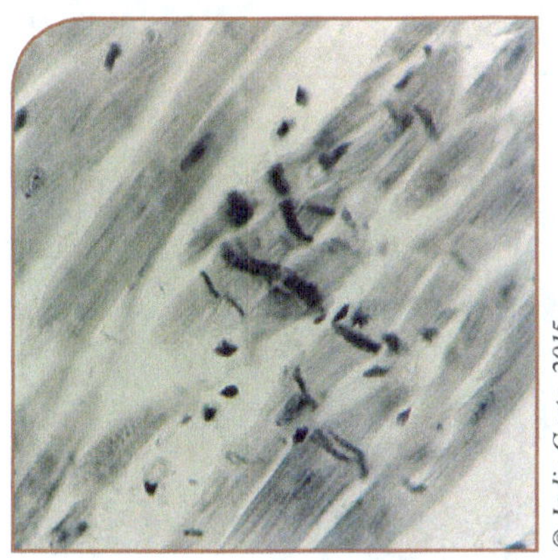
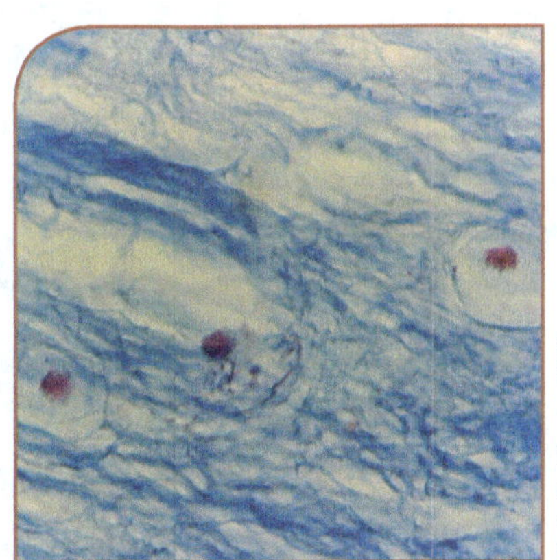
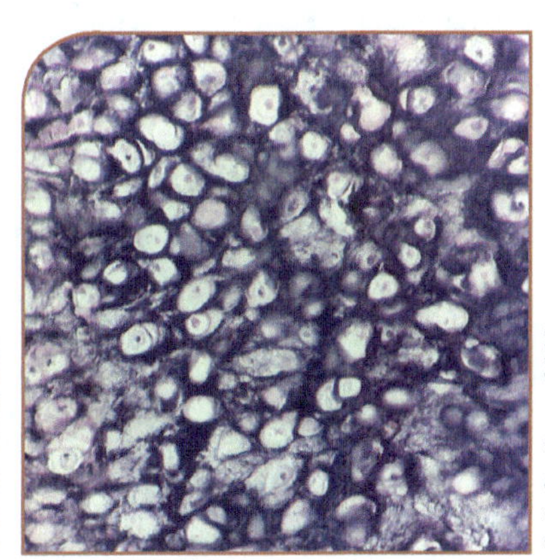

© Jodie Gerts, 2015

6. Describe the function of each type of cartilage and the predominant fiber type.

Types of Cartilage		
	Predominant Fiber Type	**Function**
Elastic Cartilage		
Fibrocartilage		
Hyaline Cartilage		

7. Shown below are parts of the cartilage tissue. Label the following parts:

Chondrocyte　　　　Lacunae　　　　Matrix　　　　Elastic fiber　　　　Collagen fiber

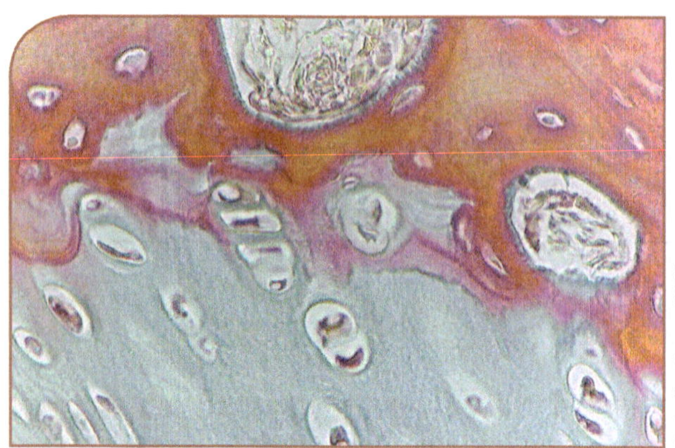

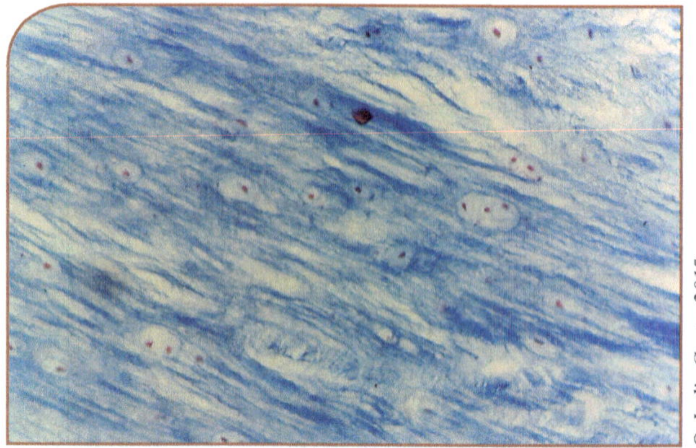

8. To the right is a slide of areolar tissue. Label the following parts:

Ground substance

Elastic fiber

Collagen fiber

Fibroblast nucleus

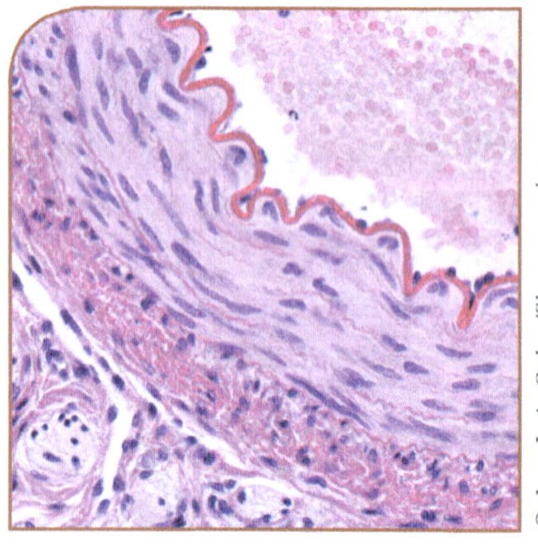

9. Shown below are the three types of muscle tissue. Identify each one.

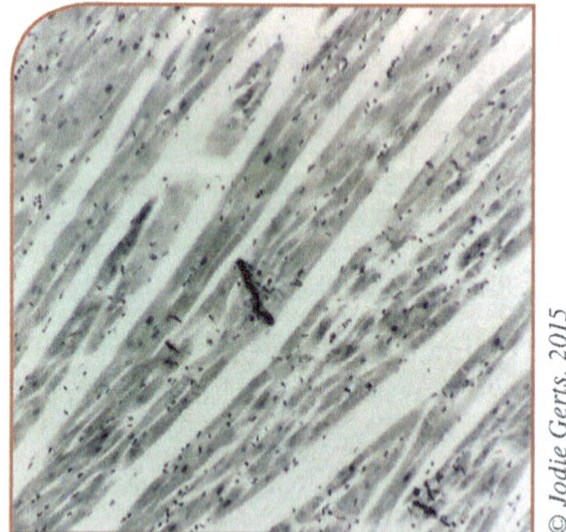

1. _______________________　　2. _______________________　　3. _______________________

A. Draw an arrow to the intercalated discs.

B. Two muscle types are striated…which are they?

_______________________________　　_______________________________

C. Where is smooth muscle found? ____

D. Skeletal muscle tissue fibers (cells) are multinucleated. What does that mean? _______________

____

10. Answer the following questions about osmosis:

A. What will happen to red blood cells when exposed to 0% saline solution? _______________

____

B. What would happen if sterile water would be given to a patient through an IV? _______________

____

C. What will happen to red blood cells when exposed to 10% saline solution? _______________

____

D. Why would red blood cells not react to the 0.9% saline solution? _______________

____

E. Is the plasma membrane of the red cells permeable to NaCl? _______________ Is it permeable to

water? _______________

F. What transport system prevents the red blood cell from swelling in a healthy individual?

____

G. You are swimming in the ocean for an hour. Draw what happens to your body in terms of osmosis.
Describe the solution (ocean water) using the terms isotonic, hypotonic, or hypertonic.

____

____

____

H. Assuming the body's cells are permeable to glucose, what happens over time if the patient is infused with a solution of 5% dextrose in normal saline (D5NS)? Note that dextrose is glucose.

____

____

____

I. You have a patient who lost one liter of blood, and you need to restore volume quickly while waiting for a blood transfusion to arrive from the blood bank. Which would be better to administer: 5% dextrose in water or 0.9% NaCl in water? (Hint: think about how these solutes distribute in the body.) Defend your answer.

____

____

____

LAB 3: TISSUES AND PROTEIN SYNTHESIS

Animations

Transcription (mRNA synthesis)

1. What does transcription synthesize? ____

2. *RNA   DNA*   polymerase binds to the promoter region of the DNA molecule.   (circle)

3. It moves along the __________________ strand of the DNA.

4. What is added to the growing strand of mRNA? _______________________________________

Translation (Protein synthesis)

1. Name the four requirements to begin translation.

    _________________________________   _________________________________

    _________________________________   _________________________________

2. Which site on the ribosome does the first tRNA bind? _________________________________

3. Which site on the ribosome does the second tRNA bind? _______________________________

4. What happens to the amino acid on the first tRNA after the second tRNA enters the ribosome?

    ____

5. How far does the ribosome shift down on the mRNA? _______________________________________

6. When does the synthesis of protein complete? _______________________________________

Endocytosis and Exocytosis

1. How do cells move material into the cell when it is too large to pass through the plasma membrane?

____

2. The three types of endocytosis are ____.

3. The largest amount of material can be brought into the cell via _________________________________.

4. Clathrin coated pits hold _________________ which are specific to individual molecules.

5. Exocytosis uses _________________ vesicles to fuse with the plasma membrane in order to move

materials _________________ of the cell.

Sodium—Potassium Pump

1. The sodium--potassium pump is an _________________ transport process which means that

_________________ is used to power the pump.

2. Active transport moves molecules _________________ the concentration gradient.

3. How many molecules of sodium are moved out of the cell? _______________________________

4. How many molecules of potassium are moved into the cell? _______________________________

Cotransport

1. Cotransport moves one molecule down its concentration gradient and one molecule _______________

its concentration gradient.

2. The type of molecule that is moved down the concentration gradient is usually _______________ or

_________________ .

3. The transmembrane protein is called a _________________ when both molecules move in the same

direction.

4. How does the sodium move back into the extracellular fluid? _______________________________

5. What powers the sodium-potassium pump? _______________________________

6. In countertransport, one ion moves down its concentration gradient and the other moves

   ___________________ its concentration gradient.

7. In countertransport, the ions move in _________________ directions.

Facilitated Diffusion

1. How do molecules move through the plasma membrane? _______________________________

2. In facilitated diffusion, molecules move   *UP   DOWN*   the concentration gradient.   (circle)

3. *TRUE   FALSE*   In order to move molecules through the plasma membrane, a cell must use ATP.   (circle)

4. What differentiates facilitated diffusion from simple diffusion? _____________________________

   ____

Module 3: Tissues

Animations

Epithelial Tissue Overview

1. Name four generalized areas where epithelial tissue is found?

   ___________________________________   ___________________________________

   ___________________________________   ___________________________________

2. The apical surface is   *free   fused*   . (circle)

3. The basal surface is   *free   fused*   . (circle)

4. Between what structures does the basement membrane lie? _______________________________

5. *TRUE   FALSE*   Epithelia lack blood vessels.   (circle)

6. *TRUE   FALSE*   Epithelia lack nerves.   (circle)

7. *TRUE   FALSE*   Epithelia have stem cells.   (circle)

8. Epithelium consisting of one layer of cells is called _________________________________.

9. Epithelium consisting of two or more layer of cells is called _____________________________.

10. When apical cells are flat, they are called _____________________________________.

11. When apical cells are square, they are called _____________________________________.

12. When apical cells are rectangular, they are called _______________________.

13. What is unique about transitional epithelium? _______________________

14. Pseudostratified epithelium has cells of varying heights but all cells touch the _______________

surface.

15. What are the two subclassifications of pseudostratified epithelium?

_______________________   _______________________

16. What is the function of microvilli? _______________________

17. Which protein is found in dead stratified squamous epithelium? _______________________

18. What are the subclassifications of stratified squamous epithelium? _______________________

19. What is the function of tight junctions? _______________________

20. Do substances that enter the body pass through the cell or between the cells? _______________________

21. Which type of intercellular junction resists mechanical stress? _______________________

22. What can pass from cell to cell through a gap junction? _______________________

Connective Tissue Overview

1. Which three components does all connective tissue share?

_______________________   _______________________   _______________________

2. What makes up the extracellular matrix? _______________________

3. What are the two subcategories of connective tissue proper?

_______________________   _______________________

4. How are the protein fibers arranged in loose connective tissue? _______________________

5. How much ground substance does dense connective tissue have? _______________________

6. What are the three types of loose connective tissue?

_______________________   _______________________   _______________________

7. A cell type that is found in areolar tissue is _______________________.

8. What is the primary cell type of adipose tissue? _______________________

9. Tendons and ligaments are made from _______________________.

10. How are the collagen fibers arranged in dense regular connective tissue? _______________________

11. What are the subdivisions of supporting connective tissue? ___________________________________

12. Cartilage's extracellular matrix functions to provide ___________________________________.

13. Bone's extracellular matrix functions to provide ___________________________________.

14. Chondrocytes are found in ___________________ within the extracellular matrix.

15. Which type of cartilage is found in the external ear? ___________________________________

16. Which type of cartilage is found in the intervertebral discs? ___________________________________

17. Which type of cartilage does not have a perichondrium? ___________________________________

18. Concentric layers of lamellae form the functional unit of compact bone called ___________________________.

19. Where are the blood vessels found in connective tissue? ___________________________________

20. Osteocytes are housed in the small spaces called ___________________________________.

21. *TRUE   FALSE*   Spongy bone contains osteons.   (circle)

22. Osteocytes are scattered in ___________________________________.

23. What is the fluid connective tissue? ___________________________________

24. Erythrocytes are also called ___________________________________.

Histology:

Simple Squamous Epithelium
   Simple squamous epithelium

Simple Columnar Epithelium (ciliated)
   Basement membrane
   Cilia on simple columnar epithelium
   Simple columnar epithelium ciliated

Pseudostratified Columnar Epithelium
   Goblet cell
   Pseudostratified columnar epithelium

Stratified Squamous Epithelium
   Stratified squamous epithelium

Transitional Epithelium
   Transitional epithelium

Areolar Connective Tissue
   Collagen fiber
   Elastic fiber
   Ground substance
   Nucleus of fibroblast

Adipose Connective Tissue
   Adipocyte

Dense Regular Connective Tissue
   Collagen fibers
   Ground substance
   Nucleus of fibroblast

Dense Irregular Connective Tissue
   Collagen fibers
   Ground substance
   Nucleus of fibroblast

Hyaline Cartilage
   Chondrocyte
   Extracellular matrix
   Lacunae

Fibrocartilage
   Chondrocyte
   Collagen fibers
   Extracellular matrix
   Lacunae

Elastic Cartilage
   Chondrocyte
   Elastic fibers
   Extracellular matrix
   Lacunae

Compact Bone
   Osteocyte in lacunae
   Osteon

Spongy Bone—know picture

Blood
   Erythrocyte
   Leukocyte

Skeletal Muscle Tissue
> Nucleus of skeletal muscle fibers

Cardiac Muscle Tissue
> Branched cardiac muscle fibers
>
> Intercalated disc

Smooth Muscle Tissue Low Magnification
> Longitudinal smooth muscle layer

Nervous Tissue
> Neuron

Microscopy

Look at the following slides under the microscope. Draw what you see in the circles provided. Be able to identify each of the following tissues and their unique structures.

Epithelial Tissues

Stratified Squamous Epithelium

> Apical surface
>
> Basal surface

Pseudostratified Columnar Epithelium w/Cilia

> Cilia
>
> Basement membrane

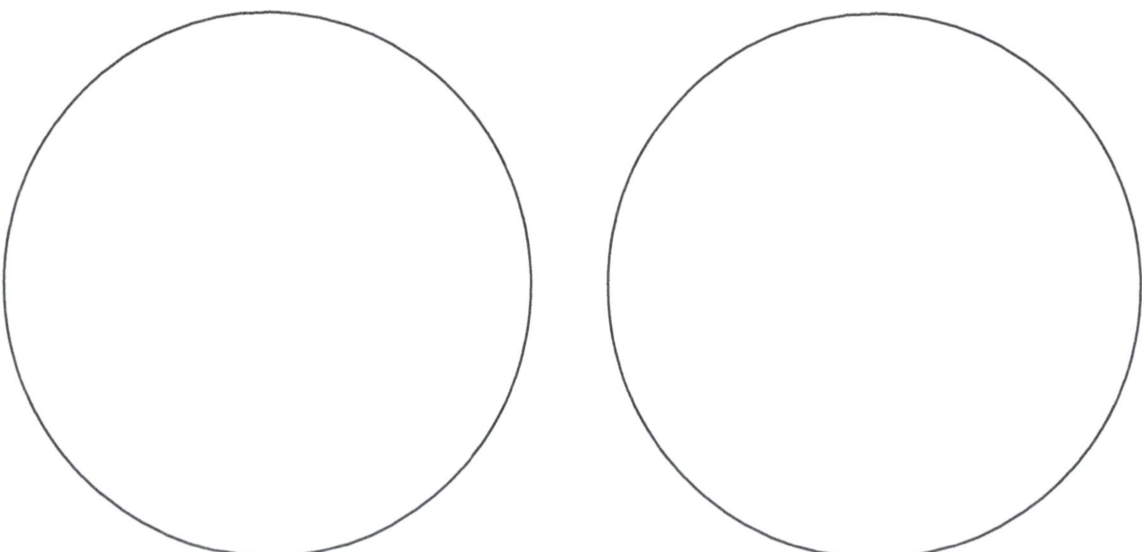

Connective Tissues

Areolar Connective Tissue

Ground substance

Collagen fibers

Elastic fibers

Fibroblasts

Dense Regular Connective Tissue

Collagen fibers: wavy, densely packed

Fibroblast nuclei may be present

Ground substance

Hyaline Cartilage

Lacunae

Chondrocyte

Extracellular matrix

Elastic Cartilage

Lacunae

Chondrocyte

Elastic fibers

Fibrocartilage

Lacunae

Chondrocytes

Collagen fibers

Blood

Erythrocytes (Red blood cells)

Leukocytes (White blood cells)

Compact Bone

Lacunae

Osteocyte

Osteon

Central canal

Muscular Tissue

Skeletal Muscle   **Smooth Muscle**

   Muscle fibers: striations   Myocytes: fusiform, nonstriated

   Multiple nuclei   Single nucleus

Red Blood CellS

Materials

Alcohol pad	1 Cover slip
Lancet (Saf-T-Pro)	1 Glass microscope slide
Gauze/Band-aid	Compound light microscope with 40x lens
Toothpick	0.9% saline solution in beaker with pipette

Procedure

1. Wipe a finger with the alcohol swab. **Twist end off lancet.** Do not pull end off lancet.
2. Quickly poke the finger with the lancet to obtain blood droplet (Use gauze/band-aid on puncture to stop bleeding).
3. Place blood into a test tube, and add several drops of isotonic saline and mix.
4. Obtain a clean microscope slide and a cover slip.
5. With pipette pointing straight down, apply one drop of blood to the center of each microscope slide.
6. Immediately mix the blood and the saline using a toothpick.
7. Gently place the cover slip over the blood/saline mixture, taking care to avoid trapping of large air bubbles underneath.
8. Observe the cell morphology under the microscope using 40x lens.
9. Dispose of the microscope slides, coverslips, toothpicks, and pipettes in the appropriate biohazard waste container.

Protein Synthesis

Directions

1. Fill in the template DNA strand from the given coding strand using the DNA base pairing rules.
2. Fill in the correct mRNA bases by transcribing the template DNA code.
3. Translate the mRNA codons to the tRNA anticodons.
4. Write in the amino acids from the Codon Table. The bases on the Codon Table are from the mRNA (not tRNA).

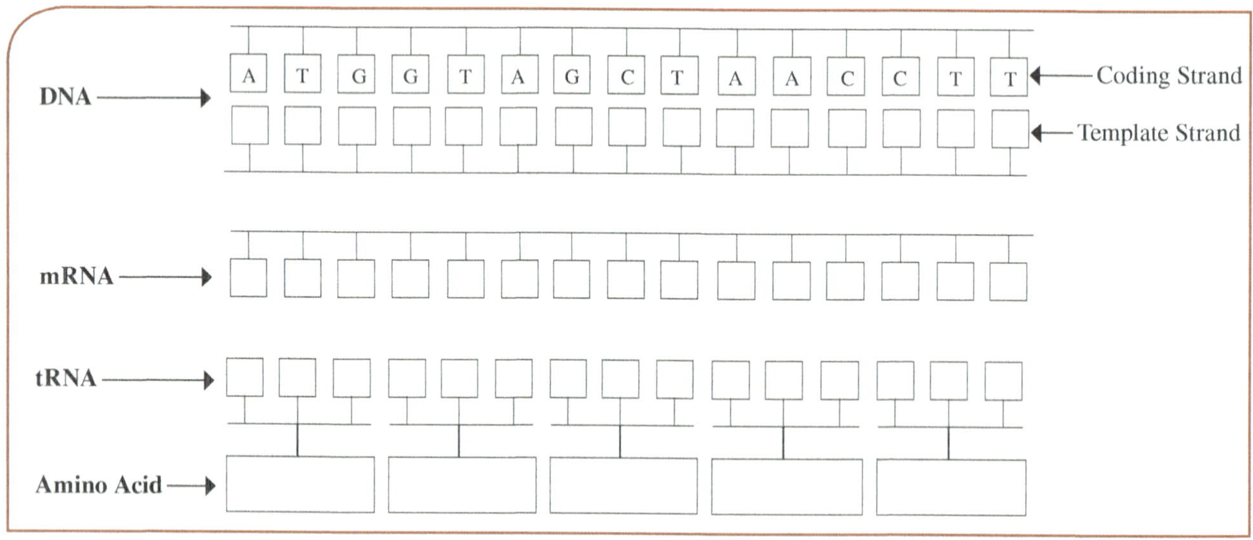

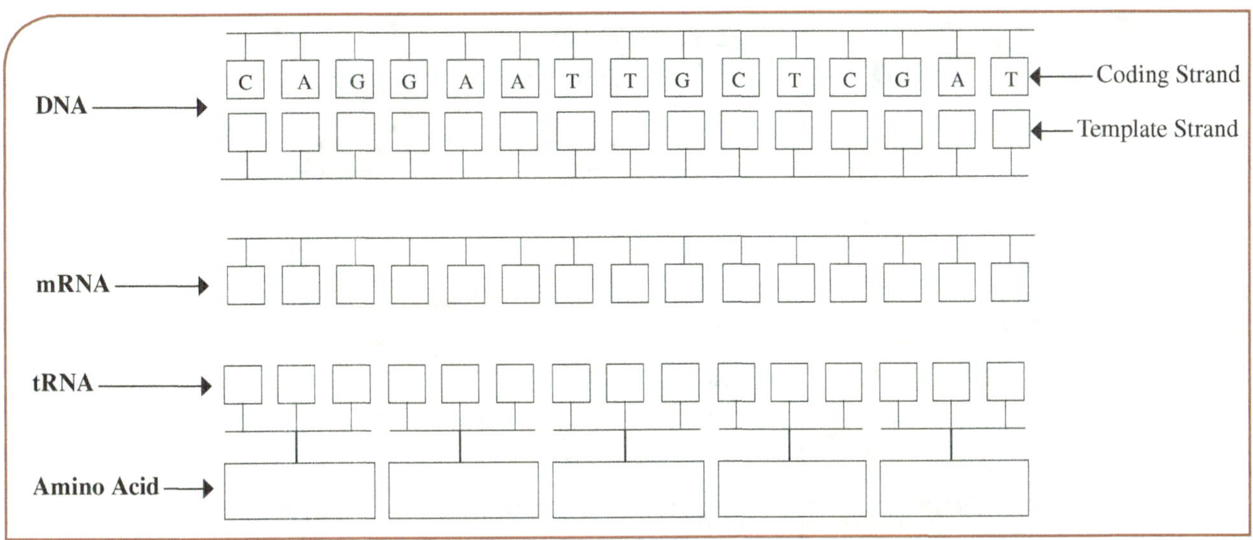

1. mRNA is synthesized in   *translation   transcription*   . (circle)

2. mRNA has   *codons   anticodons*   . (circle)

3. *1*   or   *3*   codons equal one amino acid.   (circle)

4. tRNA brings amino acids to the   *nucleus   ribosome.*   (circle)

5. A polypeptide is a sequence of   *proteins   amino acids.*   (circle)

6. tRNA has   *codons   anticodons.*   (circle)

7. tRNA transfers amino acids in   *translation   transcription.*   (circle)

Second Base

First Base	U	C	A	G	Third Base
U	UUU phenylalanine	UCU serine	UAU tyrosine	UGU cysteine	U
	UUC phenylalanine	UCC serine	UAC tyrosine	UGC cysteine	C
	UUA leucine	UCA serine	UAA stop	UGA stop	A
	UUG leucine	UCG serine	UAG stop	UGG tryptophan	G
C	CUU leucine	CCU proline	CAU histidine	CGU arginine	U
	CUC leucine	CCC proline	CAC histidine	CGC arginine	C
	CUA leucine	CCA proline	CAA glutamine	CGA arginine	A
	CUG leucine	CCG proline	CAG glutamine	CGG arginine	G
A	AUU isoleucine	ACU threonine	AAU asparagine	AGU serine	U
	AUC isoleucine	ACC threonine	AAC asparagine	AGC serine	C
	AUA isoleucine	ACA threonine	AAA lysine	AGA arginine	A
	AUG(start) methionine	ACG threonine	AAG lysine	AGG arginine	G
G	GUU valine	GCU alanine	GAU aspartate	GGU glycine	U
	GUC valine	GCC alanine	GAC aspartate	GGC glycine	C
	GUA valine	GCA alanine	GAA glutamate	GGA glycine	A
	GUG valine	GCG alanine	GAG glutamate	GGG glycine	G

Codon Table

L A B 4

Integumentary System and Skeletal System—Skull

Objectives

The purpose of this lab is to have the student understand the layers of the skin and their functions. They will be able to identify tissue types and cells within the layers using microscopy. The student will also begin the study of the skeletal system with focus on the bones and boney landmarks of the skull.

Prior to lab: Complete the Pre-lab.

Complete APR Dissection, Histology, and Animations

During lab: Complete each objective prior to leaving lab.

☐ **Integumentary Microscope Slides**
    Thick Skin
    Thin Skin

☐ **Skull Bone Identification**
    Lateral
    Superior
    Inferior
    Cranial Floor

☐ **Eye Orbit Model**

☐ **Skin Model**

LAB 4 PRELAB—Integumentary System and Skeletal System—Skull

Identify the following structures on the skin model:

Adipose tissue

Blood vessels (artery and vein)

Dermal papillae

Dermis

Epidermis

Epidermal ridges

Free nerve ending

Hair

Hair follicle

Hypodermis

Meissner's corpuscle

Nerve

Pacinian corpuscle

Sebaceous gland

Stratum basale

Stratum spinosum

Stratum granulosum

Stratum lucidum

Stratum corneum

Sweat gland

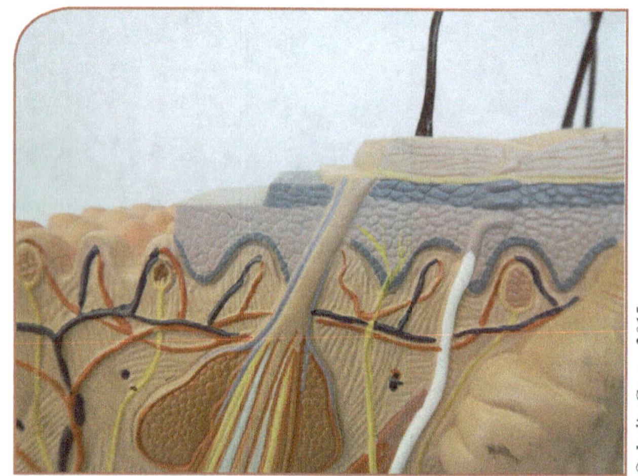

© Jodie Gerts, 2015

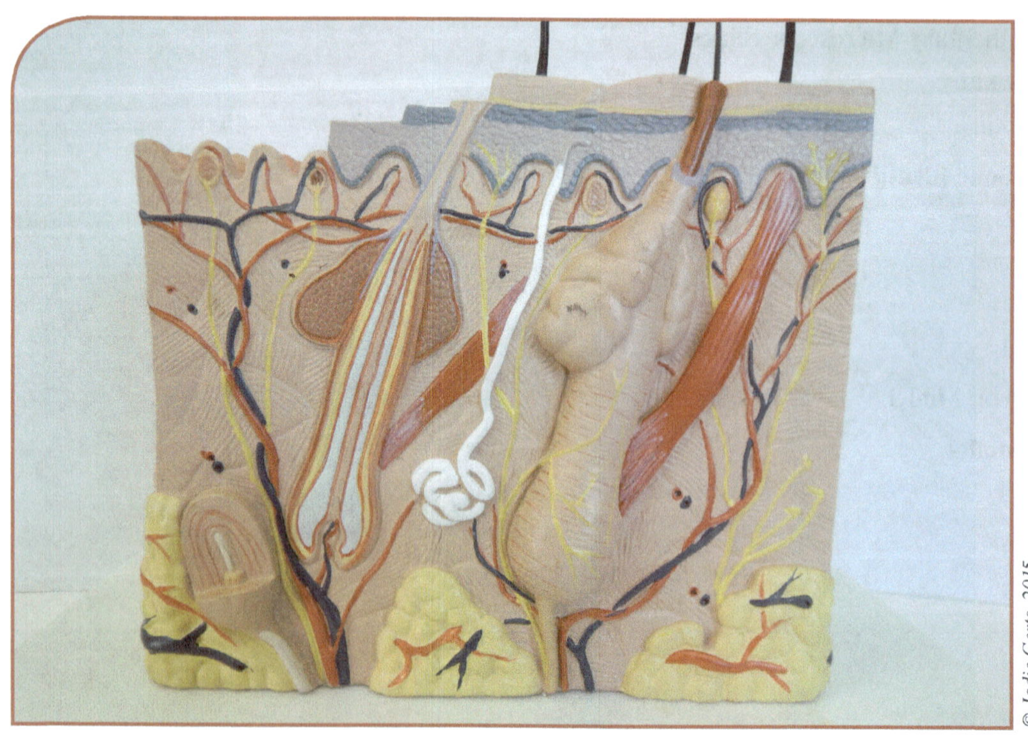

© Jodie Gerts, 2015

Label the following layers of thick skin:

Epidermis

    Stratum corneum

    Stratum lucidum

    Stratum granulosum

    Stratum spinosum

    Stratum basale

    Melanocytes

    Stem cells

    Epidermal ridge

Dermis

    Dermal papillae

    Reticular layer

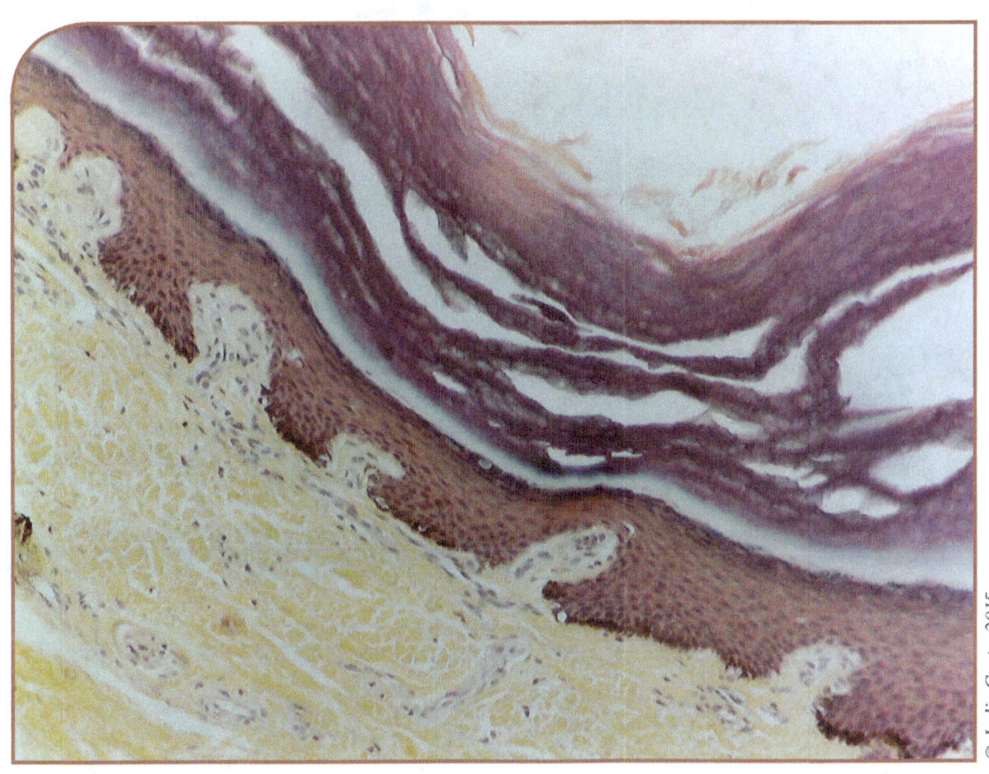

Label the following layers of thin skin:

Epidermis

    Stratum corneum

    Stratum granulosum

    Stratum spinosum

    Stratum basale

Dermis

    Reticular layer

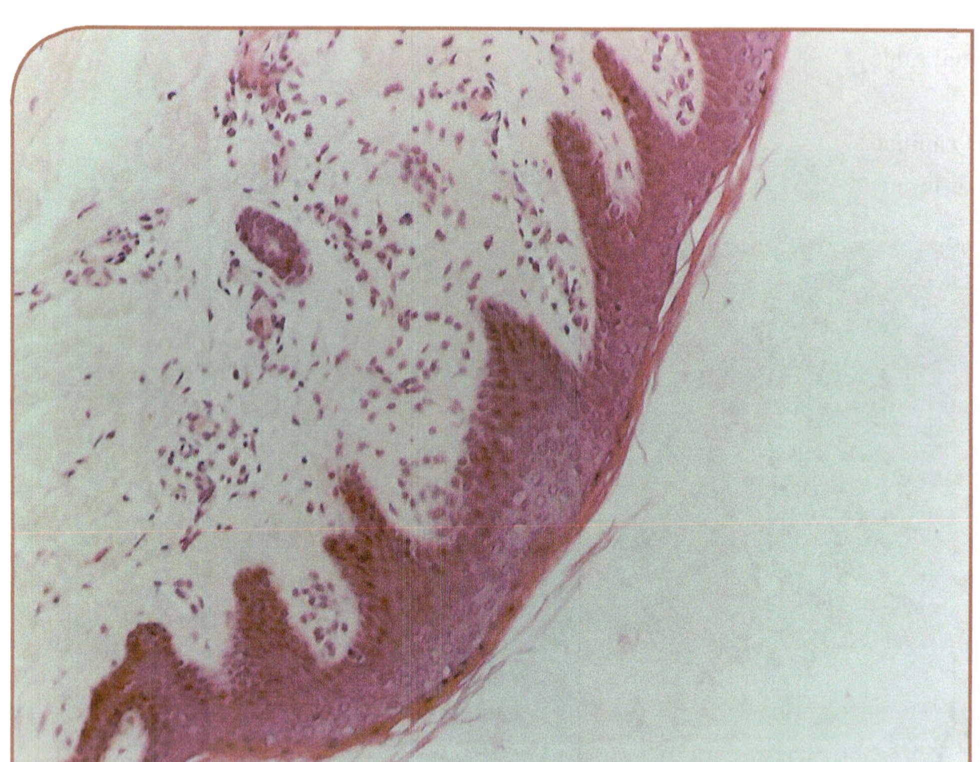

Label the following structures in the dermis of thin skin:

Hair follicle

Sebaceous gland

Arrector pili muscle

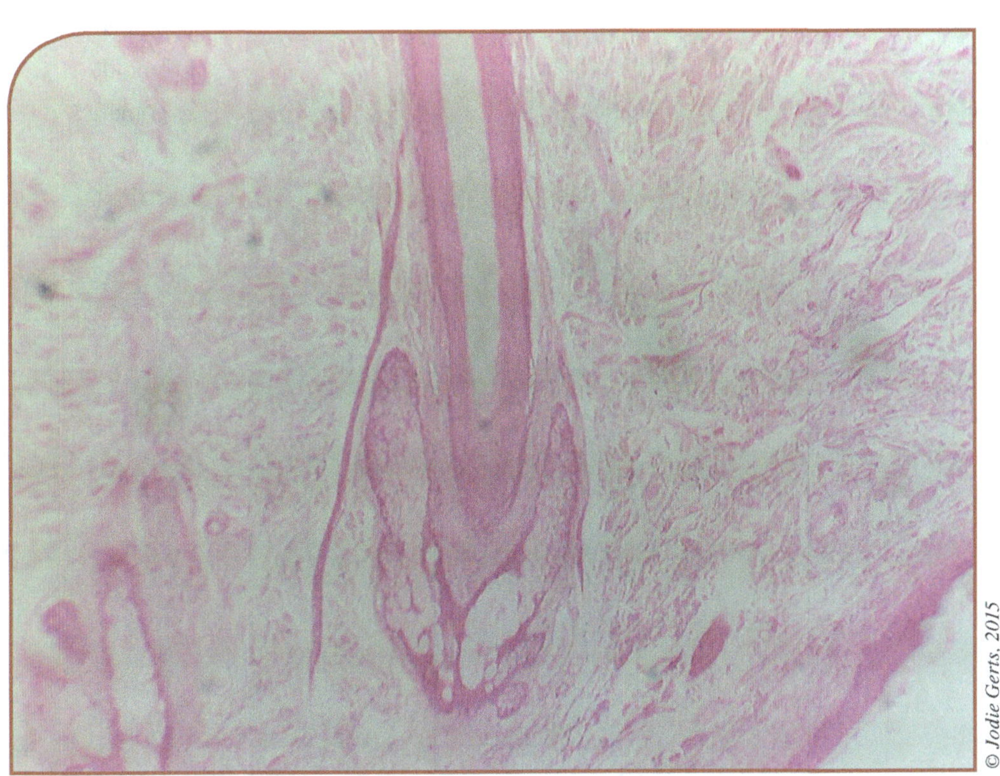

COLOR the following bones or bone markings on the skull drawings:

Anterior View	Zygomatic	Orbit
Frontal bone	Zygomatic arch	Frontal
Parietal bone	Temporal process	Maxilla
Sphenoid bone	Lacrimal bone	Sphenoid
Ethmoid bone	Nasal bone	Ethmoid
Perpendicular plate	Vomer	Zygomatic
Middle nasal conchae	Inferior nasal conchae	Palatine
Temporal Bone	Maxilla	Lacrimal
	Mandible	Lambdoid Suture

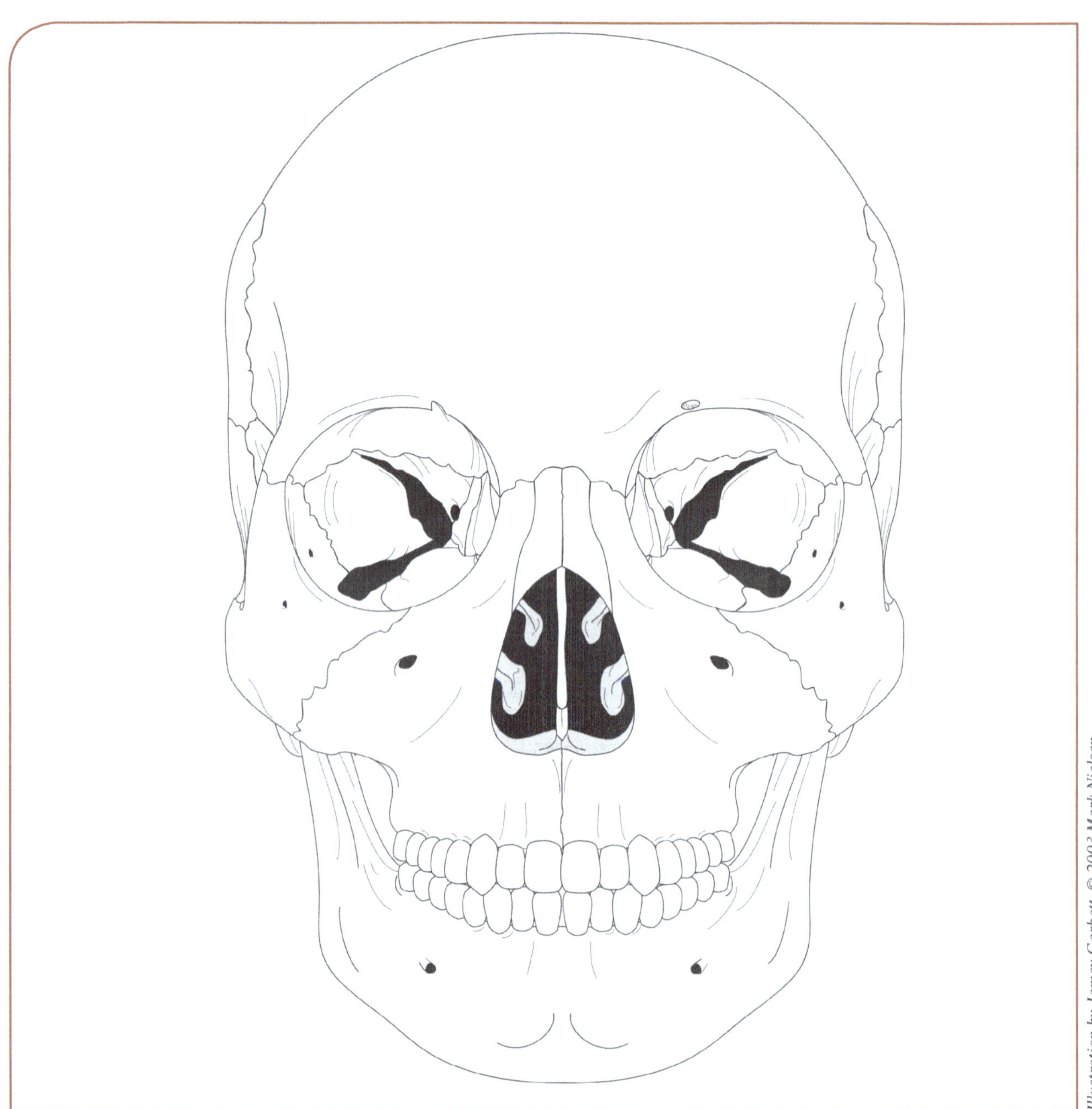

Illustration by Jamey Garbett. © 2003 Mark Nielsen

Figure 4.1 ■ Cranium, Anterior View

Lateral Views

- Frontal bone
  - Frontal sinus
- Parietal bone
- Occipital bone
- Sphenoid bone
  - Sella turcica
  - Sphenoid sinus
- Ethmoid bone
  - Perpendicular plate
  - Cribriform plate
- Temporal bone
  - Zygomatic process
  - External acoustic meatus
  - Internal acoustic meatus
  - Mastoid process
- Palatine
- Zygomatic
  - Zygomatic arch
  - Temporal process
- Lacrimal bone
- Nasal bone
- Vomer
- Maxilla
- Mandible
  - Mandibular condyle
- Jugular foramen
- Coronal suture
- Sagittal suture
- Squamous suture
- Lambdoid suture
- Frontal sinus
- Sphenoid sinus

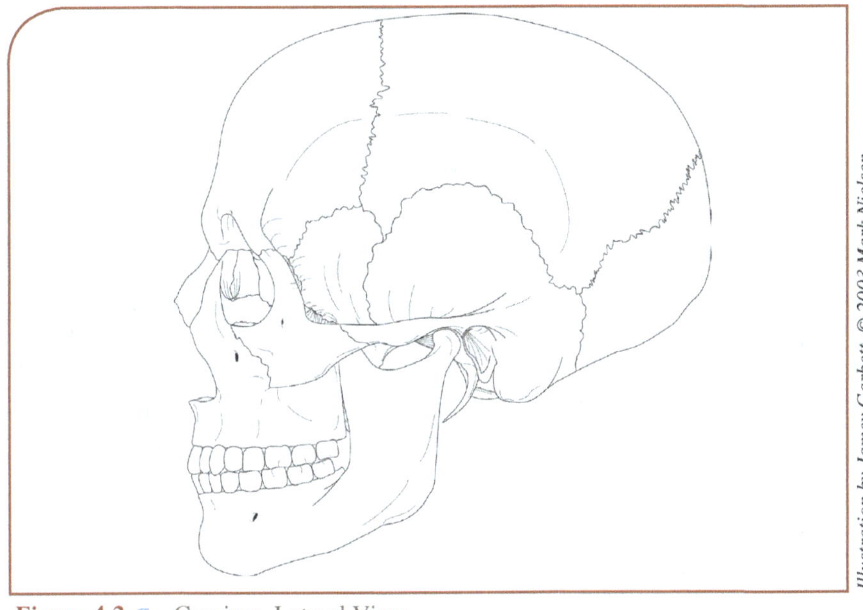

Illustration by Jamey Garbett. © 2003 Mark Nielsen

Figure 4.2 ◢ Cranium, Lateral View

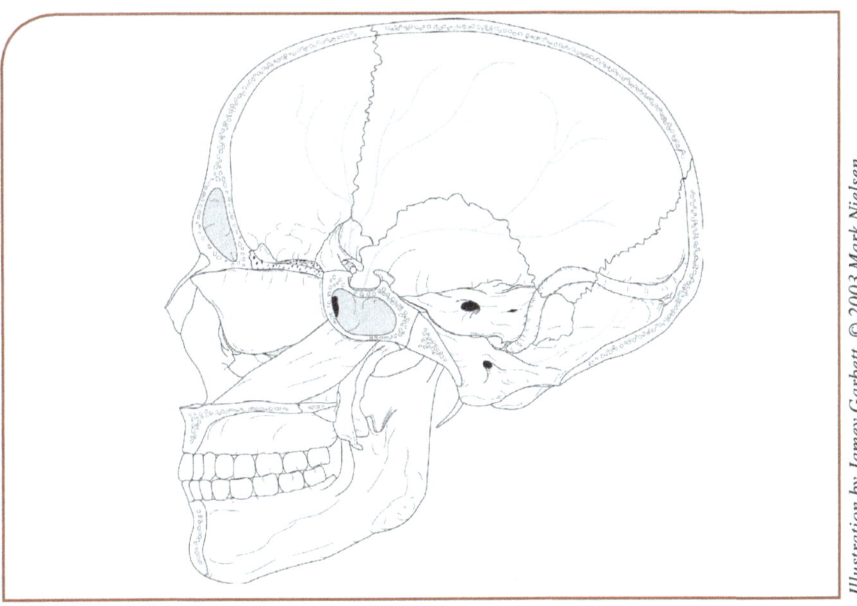

Illustration by Jamey Garbett. © 2003 Mark Nielsen

Figure 4.3 ◢ Cranium, Sagittal Section

Superior and Inferior Views

Frontal bone

Parietal bone

Occipital bone

   Occipital condyles

   Foramen magnum

Sphenoid bone

   Greater wing

   Lesser wing

   Sella turcica

Ethmoid bone

   Cribriform plate

Temporal bone

   Zygomatic process

   Mandibular fossa

   External acoustic meatus

   Internal acoustic meatus

   Mastoid process

Palatine

Zygomatic bone

Zygomatic arch

Lacrimal bone

Maxilla

Mandible

   Mandibular condyle

Jugular foramen

Anterior cranial fossa

Middle cranial fossa

Posterior cranial fossa

Coronal suture

Sagittal suture

Lambdoid suture

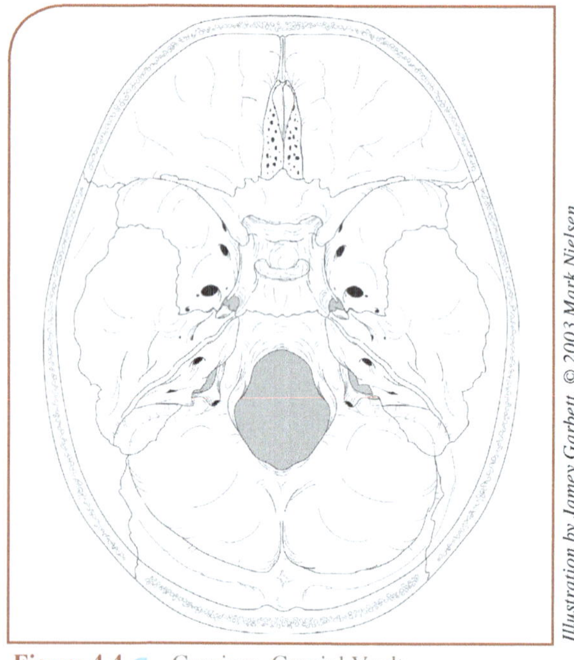

Illustration by Jamey Garbett. © 2003 Mark Nielsen

Figure 4.4 ▄ Cranium, Cranial Vault

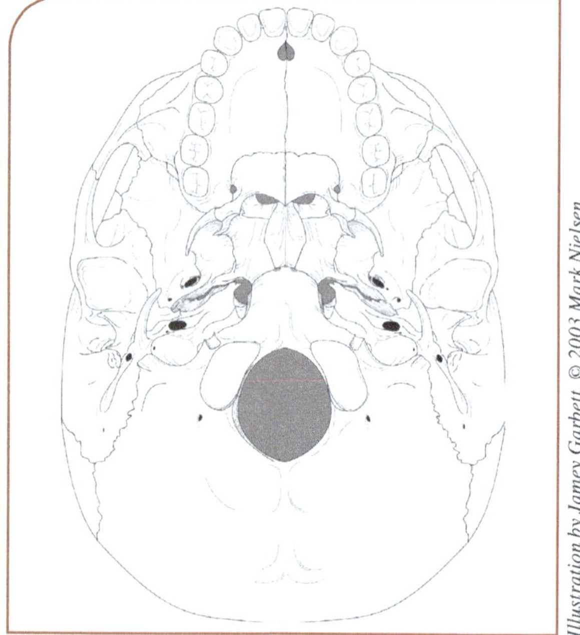

Illustration by Jamey Garbett. © 2003 Mark Nielsen

Figure 4.5 ▄ Cranium, Inferior View

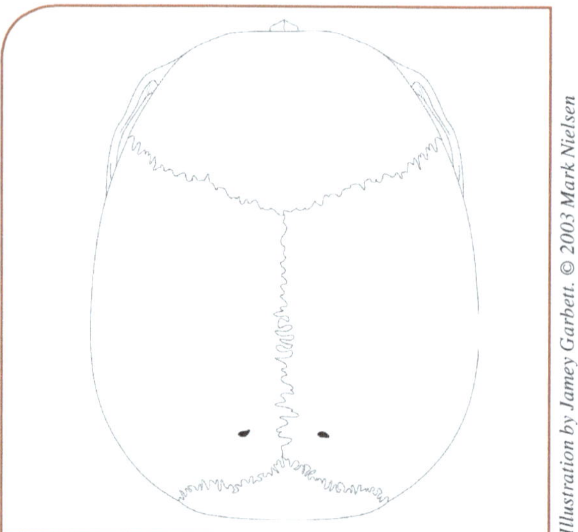

Illustration by Jamey Garbett. © 2003 Mark Nielsen

Figure 4.6 ▄ Cranium, Superior View

Define the following bone markings from your textbook or other source:

1. Condyle: ____

2. Facet: ____

3. Head: ____

4. Epicondyle: ____

5. Process: ____

6. Spine: ____

7. Trochanter: ____

8. Tubercle: ____

9. Tuberosity: ____

10. Fossa: ____

11. Foramen: ____

12. Meatus: ____

13. Sinus: ____

Skin Questions:

Name the five epidermal layers of thick skin: Name the four epidermal layers of thin skin:

_______________________________   _______________________________

_______________________________   _______________________________

_______________________________   _______________________________

_______________________________   _______________________________

____

Sensory Receptor Questions:

1. What type of sensory information does the Pacinian corpuscle detect? ___________________________

2. What type of sensory information does the free nerve ending detect? ___________________________

3. What type of sensory information does the Meissner's corpuscle detect? ___________________________

General Skin Questions:

1. Melanocytes are located in the _________________ layer.

2. Epithelial stem cells are located in the _________________ layer.

3. Langerhan's cells are located in the _________________ layer.

4. The function of Langerhan's cells is for ____ .

5. On which part of the body is thick skin located? ____

MODULE 4: INTEGUMENTARY SYSTEM

Dissection

Thin Skin and Subcutaneous Tissue

LAYER 1: Hair, epidermis

List five cells in the epidermis.

_________________   _________________   _________________

_________________   _________________

Does the epidermis have blood vessels?   *YES   NO*   (circle)

Hair is only found in   *thin   thick*   skin.   (circle)

LAYER 2: Dermis

The dermis is made of the papillary layer and the reticular layer. The papillary is primarily composed of

_________________ tissue. The reticular layer is primarily composed of _________________ tissue.

What type of muscle insert on the dermis? ____

What proteins give the dermis its strength? ____

LAYER 3: Hypodermis

What is the primary tissue type of the hypodermis? ____

What type of information does the Pacinian corpuscle receive? _________________________________

Finger Nail

LAYER 2: Eponychium, nail matrix, nail body, dermis, epidermis

What does the nail body correspond to in the skin? _______________________________________

What is the technical name for cuticle? _______________________________________

Which part of the nail contains the stem cells needed for nail growth? _______________________

What are the two functions of the epidermis? _______________________________________

Histology: Thick Skin, Low Magnification

Dermal papilla, dermis, duct of the merocrine sweat gland, epidermis, stratum basale, stratum corneum, stratum granulosum, stratum lucidum, stratum spinosum

What are the cell types of the stratum corneum? _______________________________________

Which layer has the stem cells? _______________________________________

Which layer has Langerhan's cells? _______________________________________

Which layer has keratinocytes connected by desmosomes? _______________________________

Which layer is found only in the palms of the hand and soles of the feet? ___________________

Which layers of the skin has the duct of the merocrine sweat gland? ___________________

Histology: Thick Skin, High Magnification

Dermis, epidermis, stratum basale, stratum corneum, stratum granulosum, stratum lucidum, stratum spinosum

What is the function of the stratum corneum? _______________________________________

What makes up a majority of household dust? _______________________________________

Histology: Thin Skin, Low Magnification

Dermis, duct of the sebaceous gland, epidermis, hair, hair follicle, sebaceous gland

Histology: Thin Skin, High Magnification

Dermis, epidermis, stratum basale, stratum corneum, stratum granulosum, stratum spinosum

Histology: Sebaceous Gland, Low Magnification

Arrector pili muscles, hair follicle, sebaceous gland and duct, sweat gland

What is the function of the sebaceous gland? _______________________________________

What type of muscle is the arrector pili muscle? ___________________________________

What structure is responsible for "goose bumps"? __________________________________

Histology: Merocrine Sweat Gland, Low Magnification

Dermis, epidermis, hypodermis, merocrine sweat gland

Which body parts have the highest number of sweat glands? _________________________

Which gland is responsible for temperature regulation?   *Merocrine sweat*   *Sebaceous*   (circle)

Histology: Apocrine Sweat Gland

Apocrine sweat gland

When does the apocrine sweat gland become functional? ____________________________

Where on the body is the greatest abundance of apocrine sweat glands? ______________

MODULE 5: THE SKELETAL SYSTEM

Dissection

Skull: Anterior

LAYER 2: Coronal suture, ethmoid bone, frontal bone, inferior nasal concha, lacrimal bone, mandible, maxilla, middle nasal concha, nasal bone, orbit, parietal bone, sphenoid bone, temporal bone, vomer, zygomatic bone

List the seven bones that constitute the orbit: ______________________________________

____

List the bones that form the nasal septum: ____

____

Skull: Superior

LAYER 1: Coronal suture, frontal bone, lambdoid suture, occipital bone, parietal bone, sagittal suture

The sagittal suture is between the _________________ bone and the _________________ bone.

The lambdoid suture divides the _________________ bone and the _________________ bone.

The coronal suture divides the _________________ bone and the _________________ bone.

Skull: Lateral

LAYER 2: Coronal suture, external acoustic meatus, frontal bone, greater wing of the sphenoid bone, head of the mandible, lacrimal bone, mandible, mastoid process, maxilla, nasal bone, occipital bone, parietal bone, squamous suture, temporal bone, temporal process of the zygomatic bone, zygomatic arch, zygomatic bone, zygomatic process of the temporal bone

What two bones comprise the zygomatic arch? ____

To which bone does the head of the mandible articulate? _________________________________

Where does the external acoustic meatus end? ____

Skull: Posterior

LAYER 2: External occipital protuberance, lambdoid suture, occipital bone, sagittal suture, parietal bone

The lambdoid suture is between the _________________ bones and the _________________ bone.

Skull: Mid-sagittal

LAYER 2: Bony part of nasal septum, cribriform plate, ethmoid bone, frontal sinus, internal acoustic meatus, maxilla, sphenoid bone, nasal bone, palatine bone, perpendicular plate of the ethmoid bone, sella turcica, sphenoid bone, vomer

Which bone has the internal acoustic meatus in it? ______________________________________

The sella turcica is part of the _________________ bone.

Skull: Inferior

LAYER 2: Foramen magnum, jugular foramen, mandibular fossa, maxilla, occipital bone, occipital condyle, opening of carotid canal, palatine bone, temporal process of the zygomatic bone, sphenoid bone, vomer, zygomatic process of the temporal bone

What goes through the foramen magnum? ____

Which part of the skull articulates with the first vertebrae (C1)? ____

Which vein goes through the jugular foramen? ____

Which artery goes through the carotid canal? ____

Skull: Cranial Cavity, Superior

LAYER 1: Anterior cranial fossa, cribriform plate, ethmoid bone, foramen magnum, greater wing of the sphenoid bone, jugular foramen, lesser wing of the sphenoid bone, middle cranial fossa, petrous part of temporal bone, posterior cranial fossa, sella turcica, sphenoid bone

The petrous part of the temporal bone houses the __________________ (organ of hearing) and the

__________________ (organ of equilibrium).

What part of the body lies in the sella turcica? ____

What goes through the openings of the cribriform plate? ____

Orbit

LAYER 2: Frontal bone, lacrimal bone, maxilla, ethmoid bone, palatine bone, sphenoid bone, zygomatic bone

Is the nasal bone part of the orbit?   *YES   NO*   (circle)

Ethmoid

LAYER 1: Cribriform plate, ethmoid bone, middle nasal concha, perpendicular plate

Mandible

LAYER 1: Head of mandible, mandible

Occipital Bone

LAYER 1: External occipital protuberance, foramen magnum, occipital condyle, occipital bone

Sphenoid Bone

LAYER 1: Greater wing of the sphenoid bone, lesser wing of the sphenoid bone, sella turcica, sphenoid bone

Temporal Bone

LAYER 1: External acoustic meatus, internal acoustic meatus, mandibular fossa, mastoid process, petrous part of the temporal bone, temporal bone, zygomatic process of temporal bone

Hyoid Bone

LAYER 2: Hyoid bone

To which other bones does the hyoid bone articulate with? _______________________________

Animation: Appositional Bone Growth

Osteoblasts beneath the _______________ begin the process of appositional bone growth.

What once was the periosteum now becomes the _______________ that forms around the blood vessel.

The osteoblasts form multiple layers of _______________ that become a new osteon.

Animation: Skull

The skull is made up of _______________ cranial bones and _______________ facial bones.

Other than the temporomandibular joint, all other joints in the skull are _______________ joints.

What goes through foramina? _______________________________

Which interior bone makes up the anterior base of the cranium? _______________________________

What part of the sphenoid bone houses the pituitary gland? _______________________________

What part of the ethmoid bone forms the nasal septum? _______________________________

What part of the ethmoid bone allows olfactory nerves to pass through? _______________________________

Which bone forms the posterior portion of the hard palate? _______________________________

The other bone that makes up the nasal septum is the _______________________________.

INTEGUMENTARY SLIDES

Histology of Thick Skin (10x, 40x)

1. Epidermis

    A. Stratum corneum
    B. Stratum lucidum
    C. Stratum granulosum
    D. Stratum spinosum
    E. Stratum basale
    F. Epidermal ridge

2. Dermis

    A. Dermal papillae

Histology of Thin Skin (10x, 40x)

1. Epidermis

    A. Stratum corneum
    B. Stratum granulosum
    C. Stratum spinosum
    D. Stratum basale

2. Dermis

    A. Reticular layer
    B. Hair follicle
    C. Sebaceous gland
    D. Arrector pili muscle

IDENTIFY ON SKULL MODELS

Examine lateral, superior, inferior, and the cranial floor of the skull. Be able to identify the following bones and bone markings on the skulls.

Frontal bone

Parietal bone

Occipital bone

Occipital condyles

Foramen magnum

Sphenoid bone

Greater wing

Lesser wing

Sella turcica

Ethmoid bone

Perpendicular plate

Cribriform plate

Temporal Bone

Zygomatic process

Mandibular fossa

External acoustic meatus

Mastoid process

Palatine

Zygomatic

Temporal process

Zygomatic arch

Lacrimal bone

Nasal bone

Vomer

Inferior nasal conchae

Maxilla

Mandible

Head of the mandible

Jugular foramen

Anterior cranial fossa

Middle cranial fossa

Posterior cranial fossa

Coronal suture

Sagittal suture

Squamous suture

Lambdoid Suture

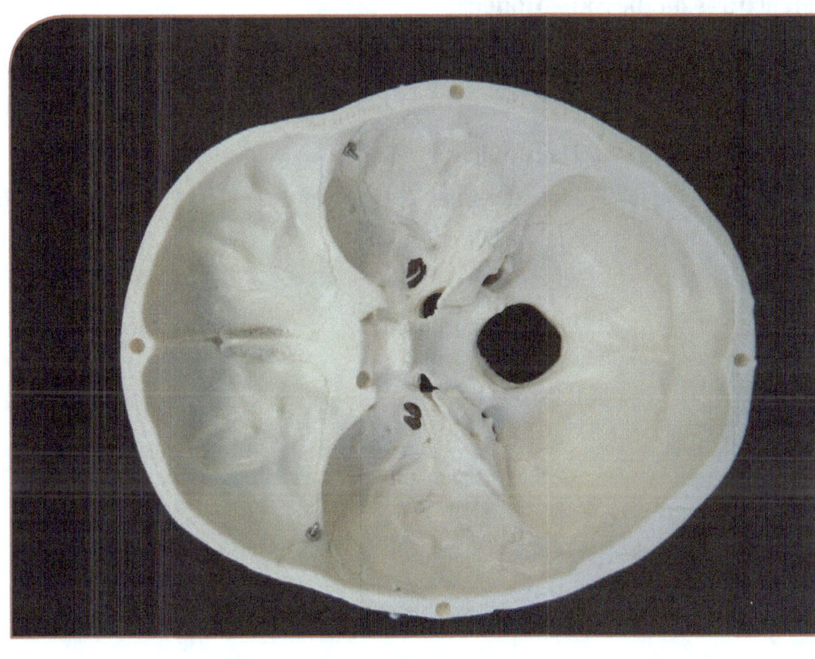

Identify the following structures on the Eye Orbit Model:

Ethmoid (7, 10)   Lacrimal (6)   Sphenoid (8)

Frontal (1)   Maxilla (4)   Zygomatic (3)

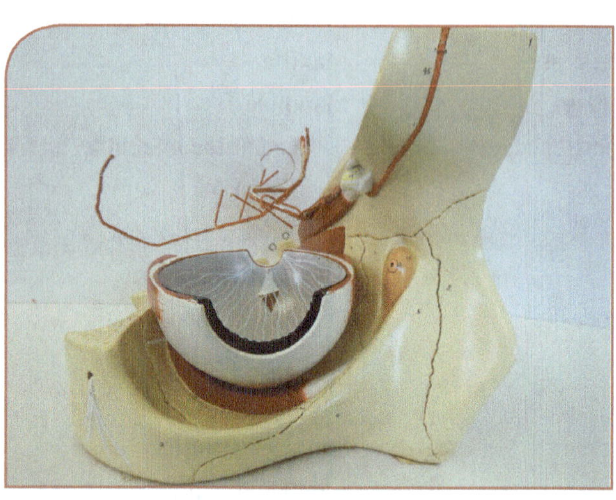
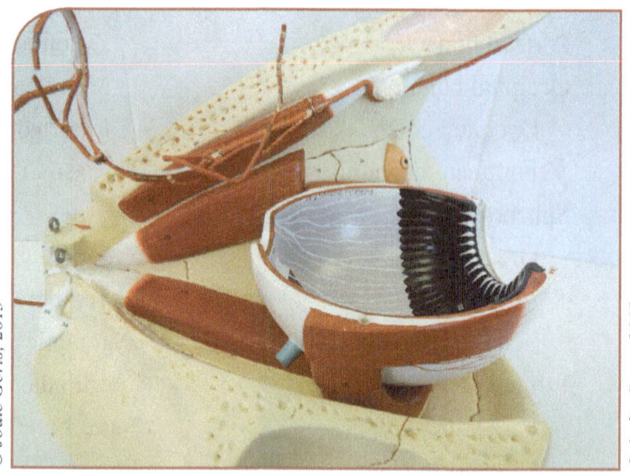

© Jodie Gerts, 2015

Identify the following structures on the Skin Model:

Adipose tissue	Free nerve ending	Sebaceous gland
Blood vessels (artery and vein)	Hair	Stratum basale
	Hair follicle	Stratum spinosum
Dermal papillae	Hypodermis	Stratum granulosum
Dermis	Meissner's corpuscle	Stratum lucidum
Epidermis	Nerve	Stratum corneum
Epidermal ridges	Pacinian corpuscle	Sweat gland

L A B *5*

The Skeletal System

Objectives

The purpose of this lab is to have the student know the bones and boney landmarks of the axial and appendicular skeleton. The student will also identify microanatomical structures of bone.

Prior to lab: Complete the Pre-lab.

Complete APR Dissection and Histology

During lab: Complete each objective prior to leaving lab.

- [ ] **Bone Microanatomy Model**
- [ ] **Bone Identification Competition**
    Thoracic Cage
    Vertebral Column
    Upper Extremity
    Pelvic Girdle
    Lower Extremity

LAB 5 PRELAB—The Skeletal System

Identify and color the following pictures from this list of bones, boney landmarks, and other structures.

Thoracic Cage

Sternum	**Ribs**
Manubrium	False ribs
Sternal body	Floating ribs
Xiphoid process	True ribs
Sternal angle	Costal cartilages
Jugular notch	Number the 12 pairs of ribs.

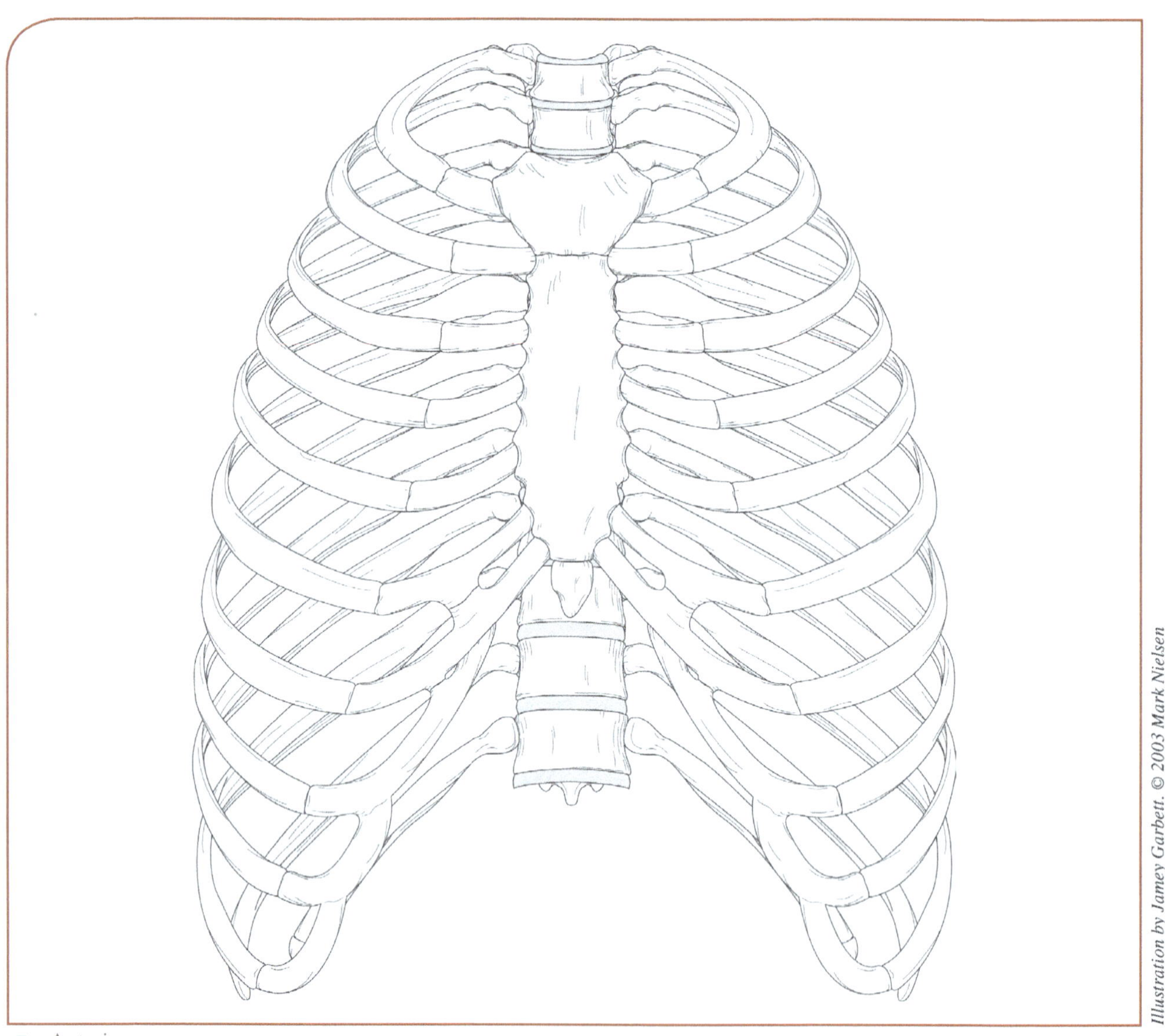

Vertebral Column

Atlas

Anterior arch

Posterior arch

Transverse foramen

Transverse process

Vertebral foramen

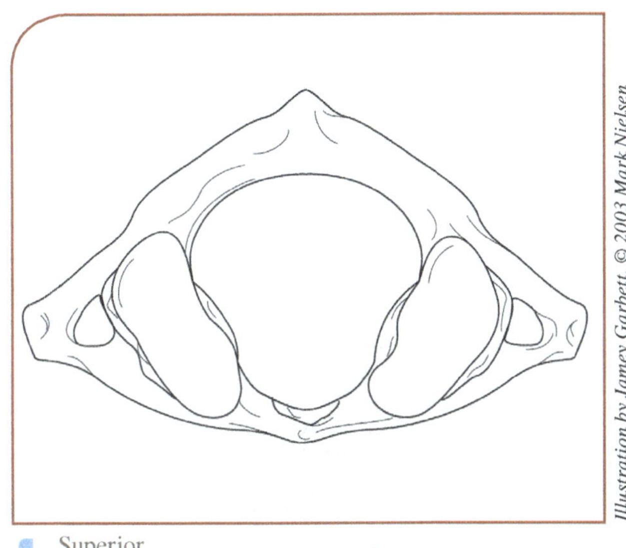

Illustration by Jamey Garbett. © 2003 Mark Nielsen

◾ Superior

Axis

Body of the axis

Dens

Inferior articular process

Lamina (difficult to see)

Transverse foramen

Vertebral foramen (unable to see)

Spinous process

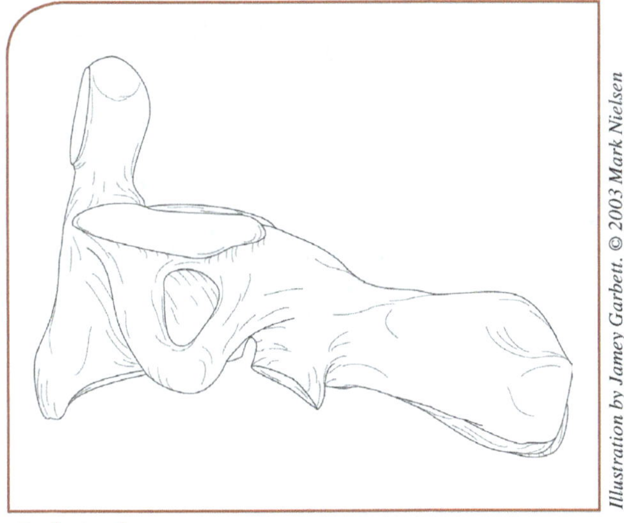

Illustration by Jamey Garbett. © 2003 Mark Nielsen

◾ Lateral

Cervical vertebra

- Body
- Inferior articular process
- Lamina
- Spinous process
- Superior articular process
- Transverse foramen
- Transverse process
- Vertebral foramen

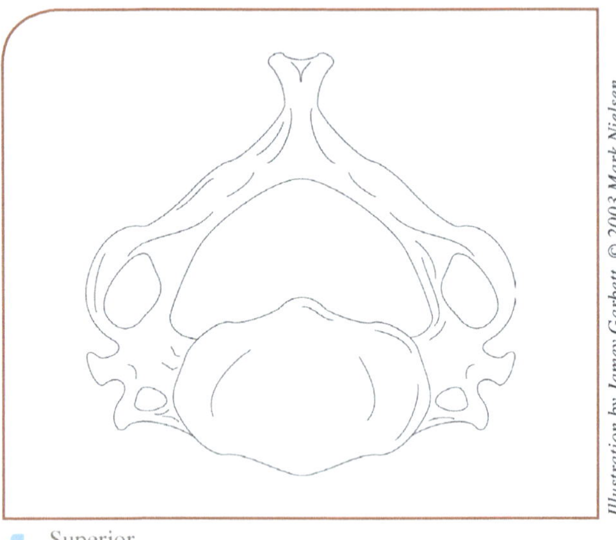

Illustration by Jamey Garbett. © 2003 Mark Nielsen

Superior

Illustration by Jamey Garbett. © 2003 Mark Nielsen

Lateral

Thoracic vertebra

- Body
- Inferior articular process
- Lamina
- Spinous process
- Superior articular process
- Transverse process
- Vertebral foramen

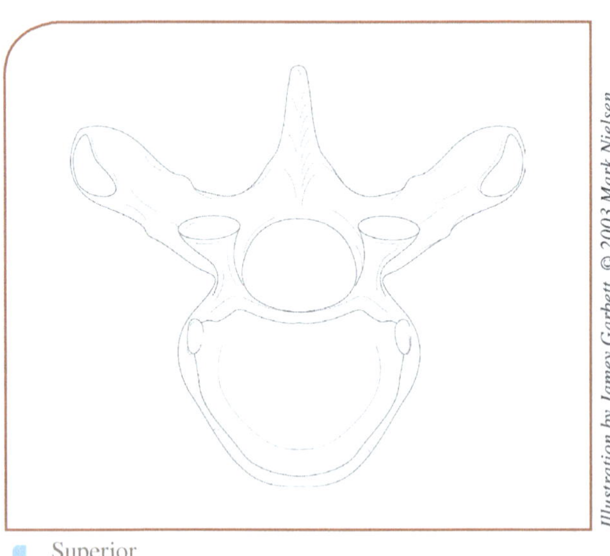

Illustration by Jamey Garbett. © 2003 Mark Nielsen

Superior

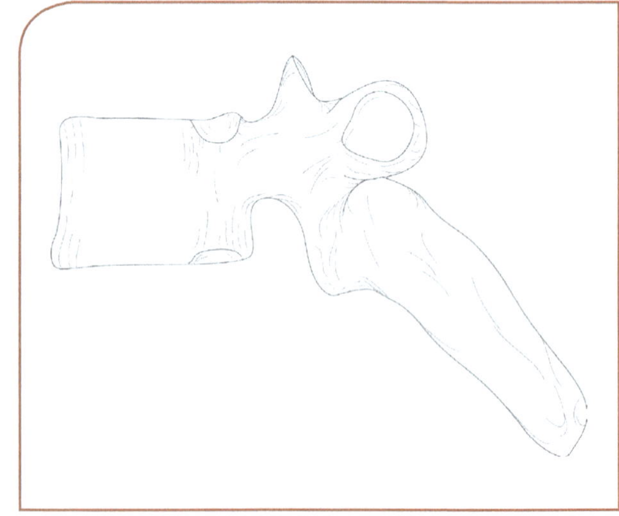

Illustration by Jamey Garbett. © 2003 Mark Nielsen

Lateral

Lumbar vertebra
- Body
- Inferior articular process
- Lamina
- Spinous process

Superior articular process
Transverse foramen
Transverse process
Vertebral formen

Superior

Lateral

Sacrum
- Superior articular process

Coccyx
Intervertebral disc
Intervertebral foramen

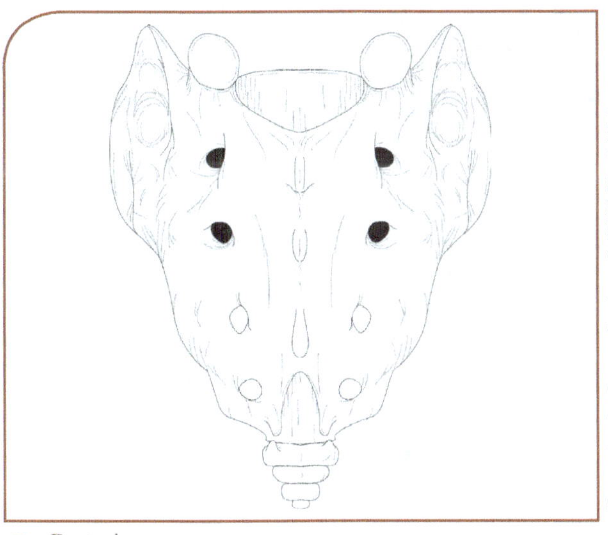

Posterior

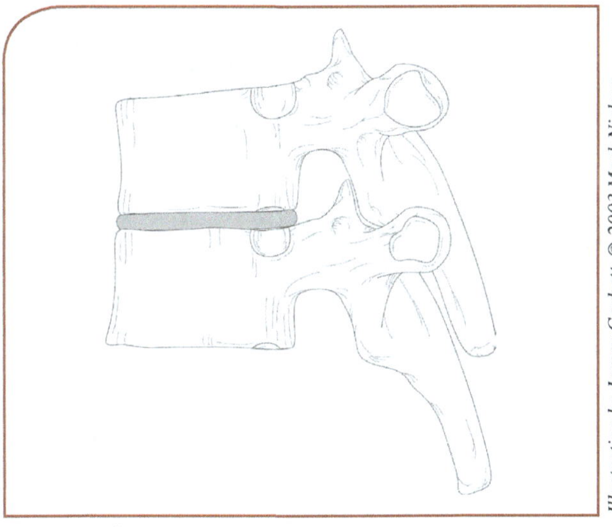

Lateral

Upper Extremity

Scapula

Acromion

Coracoid process

Glenoid cavity

Spine of the scapula

Supraspinous fossa

Infraspinous fossa

Subscapular fossa

Inferior angle

Clavicle

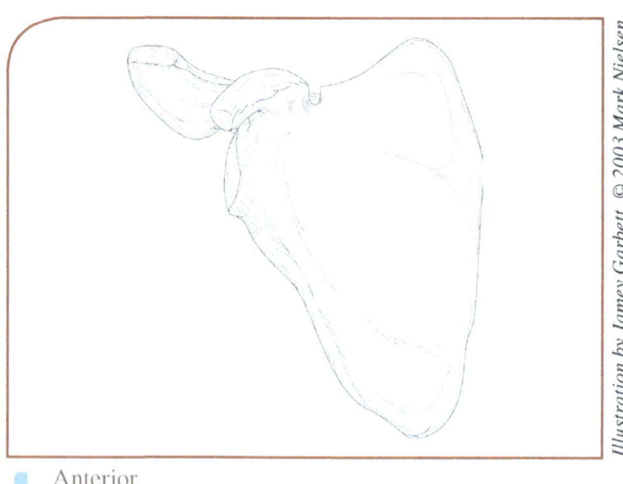

Anterior

Illustration by Jamey Garbett. © 2003 Mark Nielsen

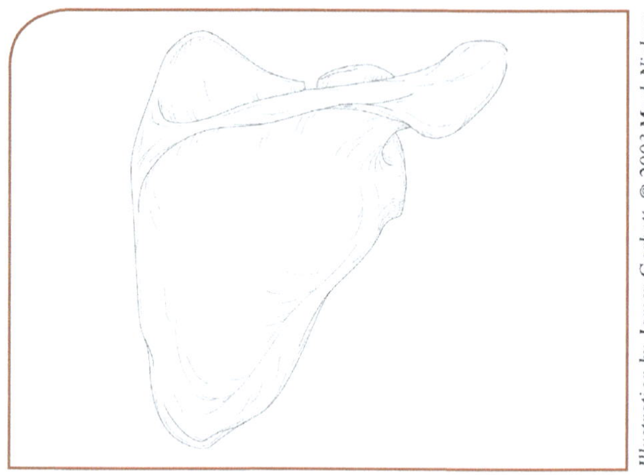

Posterior

Illustration by Jamey Garbett. © 2003 Mark Nielsen

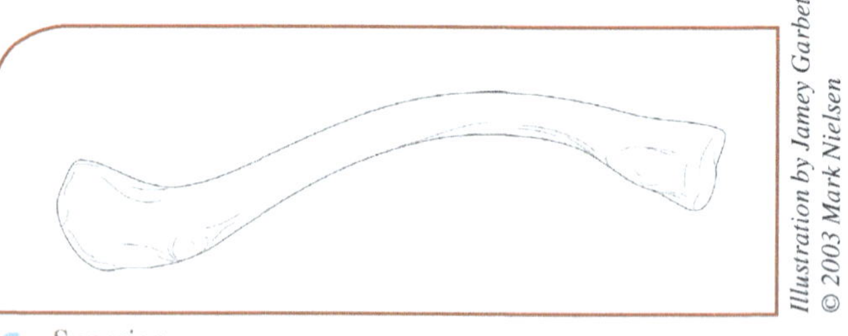

Superior

Illustration by Jamey Garbett. © 2003 Mark Nielsen

Humerus

Anatomical neck
Capitulum
Deltoid tuberosity
Greater tubercle
Head of the humerus

Lateral epicondyle
Lesser tubercle
Medial epicondyle
Surgical neck
Trochlea

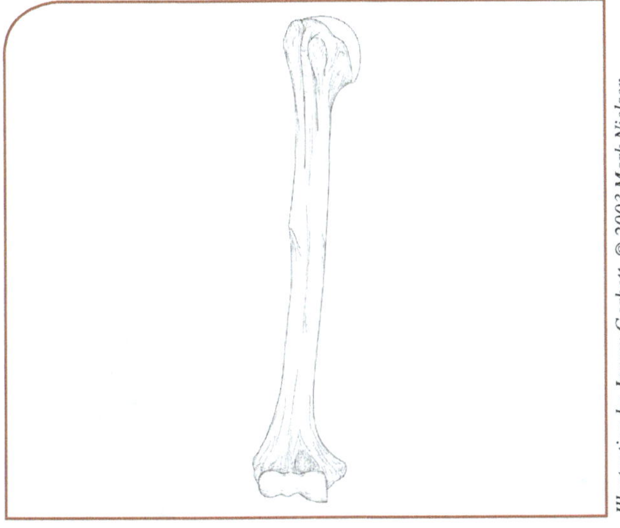

Anterior

Illustration by Jamey Garbett. © 2003 Mark Nielsen

Posterior

Illustration by Jamey Garbett. © 2003 Mark Nielsen

Ulna

Olecranon
Styloid process of the ulna
Trochlear notch

Radius

Head of radius
Radial tuberosity

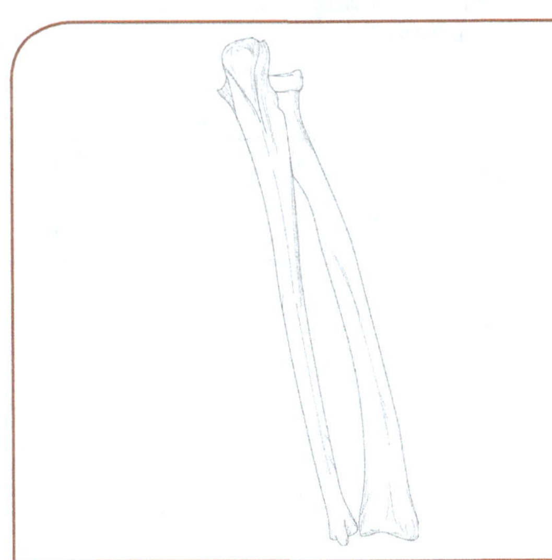

Anterior

Illustration by Jamey Garbett. © 2003 Mark Nielsen

Posterior

Illustration by Jamey Garbett. © 2003 Mark Nielsen

Carpals

Metacarpals

    Number the metacarpals:
I, II, III, IV, V

    Base of the metacarpal

    Head of the metacarpal

Phalanges (I-V)

    Proximal phalanx

    Middle phalanx

    Distal phalanx

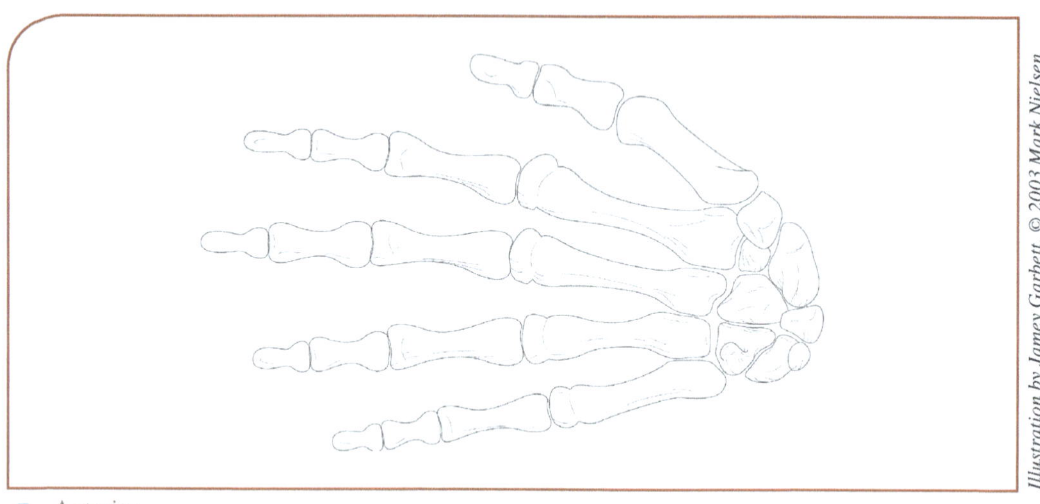

Illustration by Jamey Garbett. © 2003 Mark Nielsen

■ Anterior

Pelvic Girdle

Oscoxae or pelvic bone or
hip bone

    Acetabulum

    Ilium

        Anterior superior iliac
spine (ASIS)

Iliac crest

Iliac fossa

Posterior superior iliac
spine (PSIS)

Ischium

    Ischial tuberosity

Pubis

Pubic tubercle

    Sacroiliac joint

    Symphysis pubis

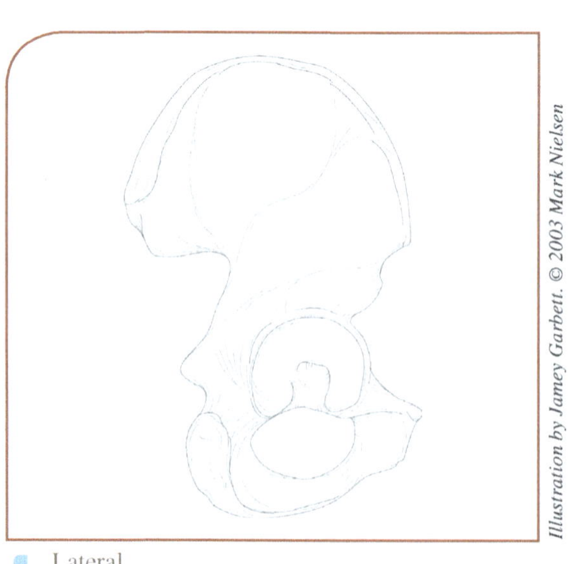

Illustration by Jamey Garbett. © 2003 Mark Nielsen

■ Lateral

Illustration by Jamey Garbett. © 2003 Mark Nielsen

■ Superior

Lower Extremity

Femur

Head of femur

Neck of femur

Greater trochanter

Lesser trochanter

Lateral condyle

Lateral epicondyle of femur

Medial condyle

Medial epicondyle of femur

Patella

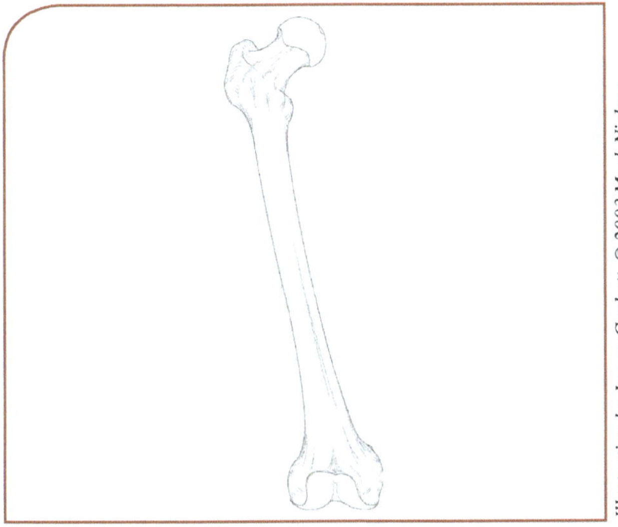

■ Anterior

Illustration by Jamey Garbett. © 2003 Mark Nielsen

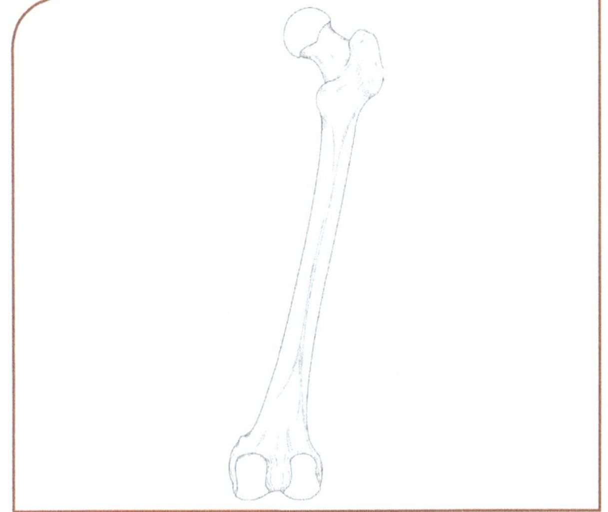

■ Posterior

Illustration by Jamey Garbett. © 2003 Mark Nielsen

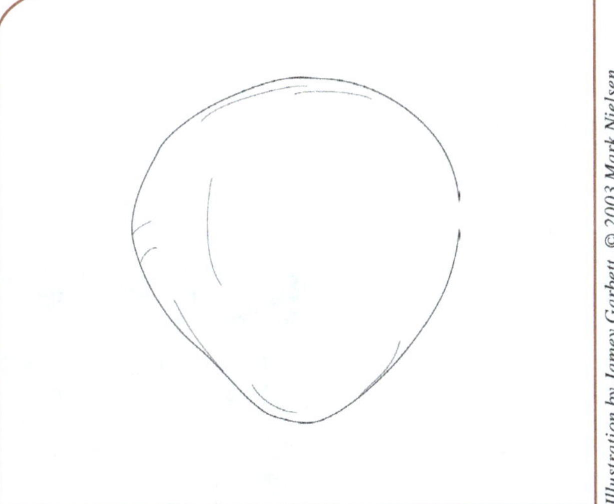

■ Anterior

Illustration by Jamey Garbett. © 2003 Mark Nielsen

Tibia
  Medial malleolus
  Tibial plateau
  Tibial tuberosity
  Head of fibula

Fibula
  Head of fibula
  Lateral malleolus
Talus
Calcaneous
Tarsals

Metatarsals
  Number the metatarsals: I, II, III, IV, V
  Head of metatarsals
Phalanges (I-V)
  Proximal phalanx
  Middle phalanx
  Distal phalanx

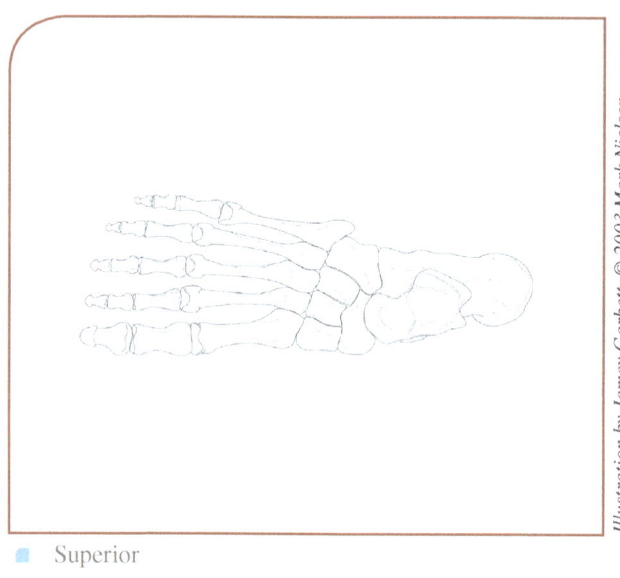

Illustration by Jamey Garbett. © 2003 Mark Nielsen

◼ Superior

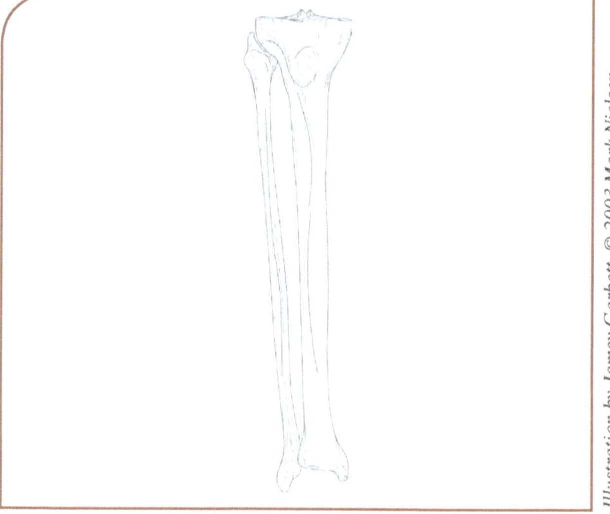

Illustration by Jamey Garbett. © 2003 Mark Nielsen

◼ Anterior

MODEL: BONE STRUCTURE MODEL

Label and identify the following structures:

1. Compact bone
2. Spongy bone
3. Periosteum
4. Endosteum
5. Bone marrow
6. Osteon
7. Interstitial lamellae
8. Concentric lamellae
9. Circumferential lamellae
10. Central canal
11. Perforating canal or Volkmann's canal
12. Sharpey's fibers
13. Trabeculae

© Jodie Gerts, 2015

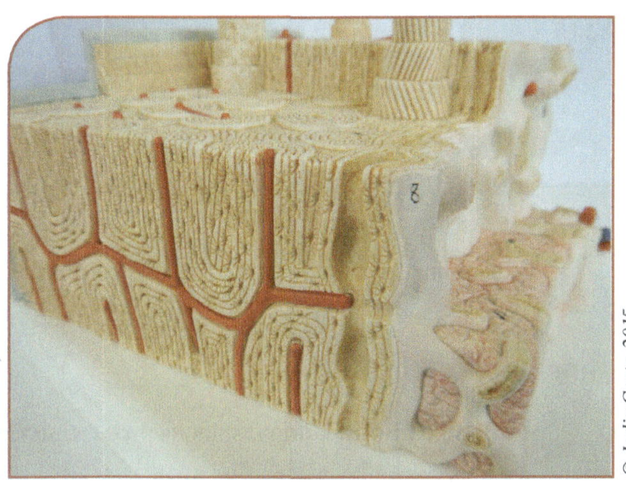

MODULE 5: SKELETAL SYSTEM

Dissection

Vertebral Column

LAYER 2: Atlas, axis, cervical vertebra, coccyx, intervertebral disc, intervertebral foramen, lumbar vertebra, sacrum, thoracic vertebra

Cervical Vertebra

LAYER 1: Body, inferior articular surface, lamina, spinous process, superior articular process, transverse foramen, transverse process, vertebral foramen

What part of the vertebrae does the spinal cord pass through? ________________________________

What passes through the transverse foramen? ________________________________

How many cervical vertebrae are there? ________________________________

Atlas

LAYER 1: Anterior arch, posterior arch, superior articular facet, transverse foramen, vertebral foramen

What does the superior articular facet articulate with? ________________________________

The atlas can also be called the ________________ vertebrae.

Is there a spinous process on this vertebra?   *YES   NO*   (circle)

Axis

LAYER 1: Body, dens, inferior articular process, lamina, superior articular process, transverse foramen, transverse process, vertebral foramen

The axis can also be called the ___________________ vertebrae.

What bone structure is unique to this vertebra? ____

Thoracic Vertebra

LAYER 1: Body, lamina, spinous process, superior articular process, transverse process, vertebral foramen

How many thoracic vertebrae are there? ____

What do the thoracic vertebrae articulate with? _______________________________________

Lumbar Vertebra

LAYER 1: Body, lamina, inferior articular process, spinous process, superior articular process, transverse process, vertebral foramen

How many lumbar vertebrae are there? ____

Which vertebrae are the largest?   *Cervical   Thoracic   Lumbar*   (circle)

Sacrum and Coccyx

LAYER 2: Coccyx, sacrum, superior articular process

The sacrum is made up of ___________________ fused vertebrae.

The coccyx is made up of ___________________ fused vertebrae.

Thoracic Cage

LAYER 2: Body of sternum, costal cartilages, false ribs, floating ribs, jugular notch of sternum, manubrium, rib 1, sternal angle, thoracic cage, true ribs, xiphoid process

There are 12 pairs of ribs, true ribs are ribs ___________________, false ribs are ribs

___________________, and floating ribs are ribs ___________________.

Circle all the parts of the thoracic cage.   *Ribs   Vertebrae   Sternum   Clavicle*   (circle)

The sternal angle articulates with the costal cartilage of _________________________________.

The most inferior aspect of the sternum is the _________________________________.

Shoulder and Arm: Anterior

LAYER 1: Acromion, anatomical neck of humerus, clavicle, glenoid cavity, head of humerus, humerus, subscapular fossa, scapula, shaft of humerus, surgical neck of humerus

The glenoid cavity of the scapula articulates with the ______________________ of the humerus.

The acromion of the scapula articulates with the _________________________________ .

Shoulder and Arm: Posterior

LAYER 1: Acromion, humerus, scapula, spine of scapula

Can you palpate the acromion?   *YES   NO*   (circle)

Can you palpate the spine of the scapula?   *YES   NO*   (circle)

Scapula

LAYER 2: Acromion, coracoid process, glenoid cavity, inferior angle of the scapula, infraspinous fossa, spine of the scapula, subscapular fossa, supraspinous fossa

Humerus

LAYER 2: Anatomical neck, capitulum, deltoid tuberosity, greater tubercle, head of humerus, lateral epicondyle of humerus, medial epicondyle of humerus, olecranon fossa, surgical neck of humerus, trochlea of humerus

Circle which boney structure is more prominent.   *Medial epicondyle   Lateral epicondyle*   (circle)

Which bone articulates with the trochlea?   *Ulna   Radius*   (circle)

Which bone articulates with the capitulum?   *Ulna   Radius*   (circle)

Which neck of the humerus is more likely to fracture?   *Surgical neck   Anatomical neck*   (circle)

Radius and Ulna

LAYER 2: Head of radius, neck of radius, olecranon, radial tuberosity, radius, styloid process of ulna, trochlear notch of ulna, ulna

The radius is the   *medial   lateral*   bone and the ulna is the   *medial   lateral*   bone.   (circle)

Wrist and Hand: Anterior and Posterior

LAYER 1: Carpal bones, distal phalanx of fingers, metacarpal I, metacarpal II, metacarpal III, metacarpal IV, metacarpal V, metacarpals, middle phalanx of fingers, proximal phalanx of fingers

Circle what bone the thumb does not have?   *Proximal phalanx*   *Middle phalanx*   *Distal phalanx*   (circle)

The pinky finger metacarpal is number ____.

Hip and Thigh: Anterior

LAYER 1: Anterior superior iliac spine (ASIS), coccyx, femur, ilium, ischium, patella, pubis, sacrum

What are the three coxal (hip) bones? _______________________________________

Hip and Thigh: Posterior

LAYER 1: Coccyx, femur, ilium, ischium, posterior superior iliac spine (PSIS), pubis, sacrum

What creates the "dimple" in the skin of the gluteal region? ___________________________

Pelvis

LAYER 1: ASIS, iliac crest, iliac fossa, ilium, pubic symphysis, pubic tubercle, pubis, sacroiliac joint, sacrum

Pelvic Girdle—Female

LAYER 2: Acetabulum, ASIS, hip bone, iliac crest, iliac fossa, ischium, pelvic girdle, pubic symphysis, pubis

What is the difference between the hip bone and the pelvic girdle? ___________________

Hip Bone

LAYER 2: Acetabulum, ASIS, iliac crest, iliac fossa, ilium, ischial tuberosity, ischium, PSIS, pubic tubercle, pubis

What part of the ischium do we sit on? ___________________________________

Femur: Anterior and Posterior

LAYER 2: Greater trochanter, head of femur, lateral condyle of femur, lateral epicondyle of femur, lesser trochanter, medial condyle of femur, medial epicondyle of femur, neck of femur, patellar surface, shaft of femur

Leg and Foot: Anterior

LAYER 1: Fibula, patella, talus, tibia, metatarsals, phalanges of toes

The tibia is the *medial   lateral* bone and the fibula is the *medial   lateral* bone. (circle)

To which bone do the inferior tibia and fibula articulate? ___________________________________

Tibia and Fibula

LAYER 2: Fibula, head of fibula, lateral condyle of tibia, lateral malleolus, medial condyle of tibia, medial malleolus, tibia, tibial tuberosity

Ankle and Foot

LAYER 2: Calcaneous, distal phalanx of toes, metatarsal I, metatarsal II, metatarsal III, metatarsal IV, metatarsal V, metatarsals, middle phalanx of toes, phalanges of toes, proximal phalanx of toes, talus, tarsal bones

Histology: Compact Bone: LM Low Magnification

Central canal of osteon, interstitial lamella, osteon

Histology: Compact Bone: SEM Low Magnification

Central canal of osteon, concentric lamella, lacuna, osteon

What type of lamellae is between osteons? *Concentric   Interstitial*   (circle)

Histology: Spongy Bone: Low Magnification

Marrow cavity of spongy bone, periosteum, trabecula

What are the predominant cells at the border of the periosteum? ___________________________________

What do those cells differentiate to? ___________________________________

Trabeculae are found in compact bone. *TRUE   FALSE*   (circle)

LAB ACTIVITIES

You will be assigned a lab partner to make a team. Using your drawings from the pre-lab, identify the bones and boney structures on the bones placed in front of you. This is a competition! Whoever identifies the bones correctly in the fastest amount of time will win a prize. Knowing your bones ahead of time will be important prior to entering this lab.

MODEL: BONE MICROANATOMY

1. Compact bone
2. Spongy bone
3. Periosteum
4. Endosteum
5. Bone marrow
6. Osteon
7. Interstitial lamellae
8. Concentric lamellae
9. Circumferential lamellae
10. Central canal
11. Perforating canal or Volkmann's canal
12. Sharpey's fibers
13. Trabeculae

▊ L A B *6*

The Articular System

Objectives

The purpose of this lab is to have the student to have a comprehensive understanding of the articular system including being able to functionally test the integrity of knee and ankle ligaments. The student will be able to identify the joints in the body. The student will be able to know the components of synovial joints and their associated function including the name and location of individual ligaments The student will be able to identify and name osteokinematic movements. The student will be able to differentiate between subluxations and dislocations.

Prior to lab: Complete the Pre-lab.

Watch the videos.

Complete APR Dissection, Histology, and Animations

During lab: Complete each objective prior to leaving lab.

☐ **Discuss on Subluxation vs. Dislocation**

☐ **Differentiate Between Genu Varus or Genu Valgus**

☐ **Identify Joints on the Skeleton**

☐ **Build the Acromioclavicular Ligaments and Elbow Ligaments**
   Coracoacromial, Coracoclavicular, and
      Acromioclavicular Ligaments
   Glenohumeral Labrum
   MCL, LCL, Anular Ligaments

☐ **Build the Hip, Knee and Talocrural Ligaments**
   Medial and Lateral Meniscus
   ACL, PCL, MCL, LCL Ligaments
   Anterior Talofibular, Calcaneofibular,
   Deltoid Ligaments

☐ **Perform Special Tests of the Knee and Ankle with a Partner**
   Lachman's Test
   Posterior Drawer
   Valgus and Varus Tests
   Anterior Drawer

LAB 6 PRELAB—The Articular System

PART 1:

Color the joints and label with words on the skeletons. You may use the acronym in parenthesis to label.

Acromioclavicular joint (AC joint)

Carpometacarpal joints (CMC)

Cervical facet joints (i.e. C5/C6)

Distal interphalangeal joint (DIP)

Glenohumeral (GH joint)

Humeroulnar and humeroradial (elbow joint)

Hip (coxal)

Intercarpal joints

Knee joint

Lumbar facet joints (i.e. L2/L3)

Metacarpophalangeal (MCP)

Metatarsophalangeal joint (MTP)

Patellofemoral joint

Proximal interphalangeal joint (PIP)

Pubic symphysis

Radioulnar joints (proximal and distal)

Sacroiliac joint (SI joint)

Scapulothoracic (functional joint—not a true synovial joint)

Sternoclavicular joint (SC joint)

Talocrural and subtalar joints

Temporomandibular joint (TMJ)

Wrist joint (radiocarpal joint)

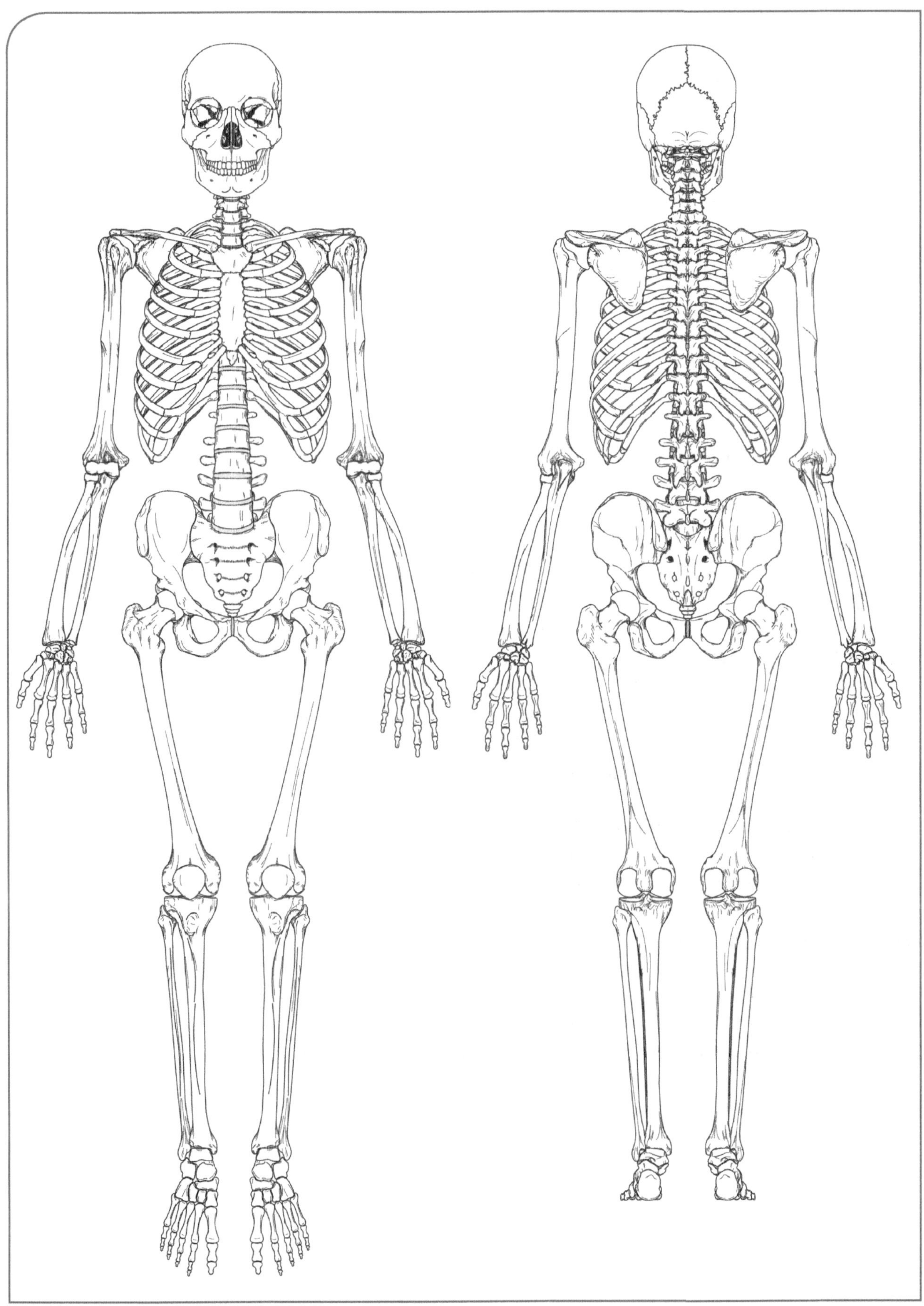

Skeleton, Anterior View Skeleton, Posterior View

PART 2:

We will be using joint motions from the following list. Below is a list of the joints we will study and motions each joint performs. NOTE: This list only contains the primary motions at each joint.

Flexion/Extension (flex/ext)

Abduction/Adduction (abd/add)

Internal rotation (Med Rot)/External rotation (Lat Rot) (IR/ER)

Supination/Pronation (sup/pro)

Right and left lateral flexion (R and L Lat flex)

Opposition (opp)

Elevation/Depression (elev/dep)

Dorsiflexion/Plantarflexion (DF/PF)

Inversion/Eversion (inv/ever)

Protraction / Retraction (prot/retr)

Right and left rotation (R and L rot)

Joint

1. **Glenohumeral**   *Triaxial*
Flexion/Extension   Abduction/Adduction   IR/ER

2. **Hip (coxal)**   *Triaxial*
Flexion/Extension   Abduction/Adduction   IR/ER

3. **Cervical facet joints, lumbar facet joints, thoracic facet joints**   *Multiaxial*
Flexion/Extension   Right and left rotation   Right and left lateral flexion

4. **Talocrural and subtalar joints**   *Multiaxial*
Dorsiflexion/Plantarflexion   Inversion/Eversion

5. **Temporomandibular joint (TMJ)**   *Multiaxial*
Elevation/Depression   Protraction/Retraction

6. **Carpometacarpal joint of thumb (CMC)**   *Biaxial*
Flexion/Extension   Opposition

7. **Metacarpophalangeal (MCP) and metatarsophalangeal joint (MTP)**   *Biaxial*
Flexion/Extension   Abduction/Adduction

8. **Scapulothoracic (functional joint)**
Elevation/Depression   Protraction/Retraction

9. **Wrist joint (radiocarpal joint)**   *Biaxial*
Flexion/Extension   Abduction/Adduction

10. **Distal interphalangeal joint (DIP) and proximal interphalangeal joint (PIP)**   *Uniaxial*
Flexion/Extension

11. **Humeroulnar and humeroradial (elbow joint)**   *Uniaxial*
Flexion/Extension

12. **Knee**   *Uniaxial*
Flexion/Extension

13. **Radioulnar joints (proximal and distal)**   *Uniaxial*
Supination/Pronation

In each picture, circle the correct positions at the listed joint

Picture 1

<u>Glenohumeral</u>:	Flex	Ext	Abd	Add
	IR	ER		
<u>Elbow</u>:	Flex	Ext		
<u>Radioulnar</u>:	Pro	Sup		
<u>Wrist</u>:	Flex	Ext		
<u>Metacarpophalangeal</u>:	Flex	Ext	Add	Abd
<u>Distal interphalangeal</u>:	Flex	Ext		

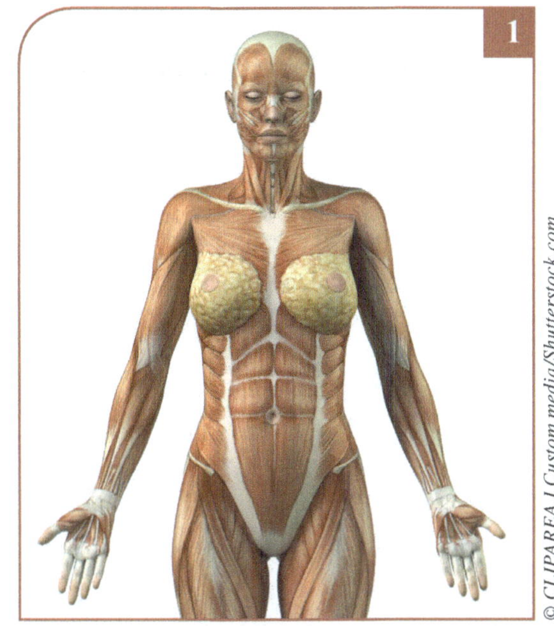

Picture 2

Right Lower Extremity

<u>Hip</u>:	Flex	Ext	Abd	Add
	IR	ER		
<u>Knee</u>:	Flex	Ext		
<u>Talocrural (ankle)</u>:	DF	PF		
<u>Metatarsophalangeal</u>:	Flex	Ext	Add	Abd
<u>Distal interphalangeal</u>:	Flex	Ext		

Left Lower Extremity

<u>Hip</u>:	Flex	Ext	Abd	Add
	IR	ER		
<u>Knee</u>:	Flex	Ext		
<u>Talocrural (ankle)</u>:	DF	PF		
<u>Metatarsophalangeal</u>:	Flex	Ext	Add	Abd
<u>Distal interphalangeal</u>:	Flex	Ext		

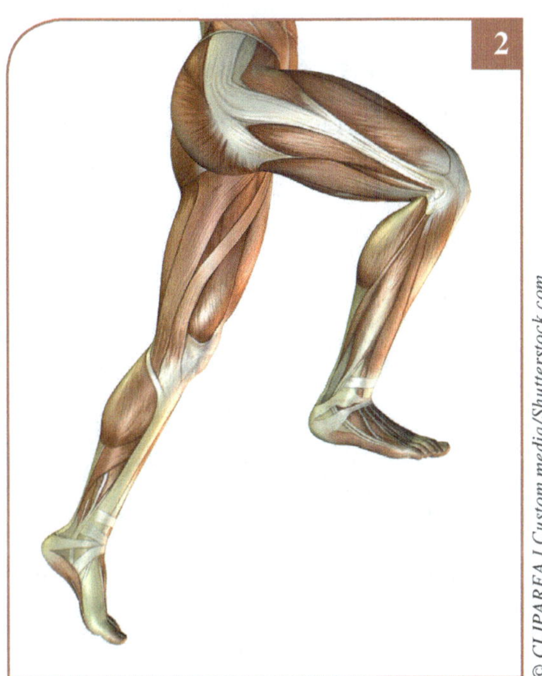

Picture 3

Glenohumeral:	Flex	Ext	Abd	Add
	IR	ER		
Elbow:	Flex	Ext		
Radioulnar:	Pro	Sup		
Metacarpophalangeal:	Flex	Ext	Add	Abd
Distal interphalangeal:	Flex	Ext		

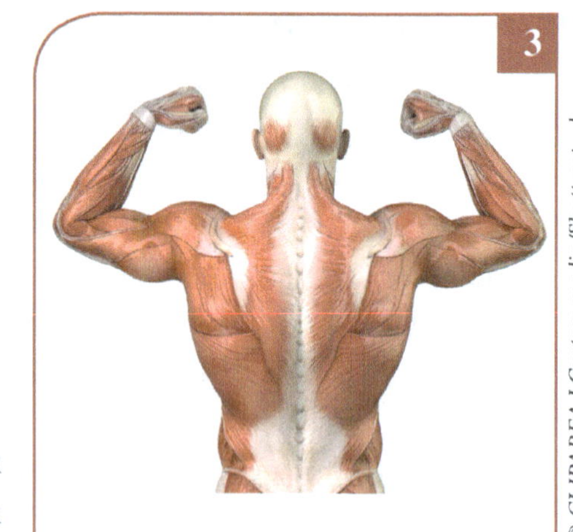

Picture 4

Cervical:	R rot	L rot	R lat flex	L lat flex
Lumbar:	Flex	Ext	R lat flex	L lat flex

Right Upper Extremity

Glenohumeral:	Flex	Ext	Abd	Add
	IR	ER		
Elbow:	Flex	Ext		
Radioulnar:	Pro	Sup		

Left Upper Extremity

Glenohumeral:	Flex	Ext	Abd	Add
	IR	ER		
Elbow:	Flex	Ext		
Radioulnar:	Pro	Sup		
Knee:	Flex	Ext		

Left Lower Extremity

Hip:	Flex	Ext	Abd	Add
	IR	ER		

Right Lower Extremity

Hip:	Flex	Ext	Abd	Add
	IR	ER		
Knee:	Flex	Ext		

Picture 5

Cervical:	Flex	Ext	R rot	L rot
	R lat flex	L lat flex		
Lumbar:	Flex	Ext	R rot	L rot
	R lat flex	L lat flex		
Scapulothoracic:	Prot	Retr	Elev	Dep
Glenohumeral:	Flex	Ext	Abd	Add
	IR	ER		

Answer the following questions about joint motions.

1. Flexion of the elbow occurs in the *frontal sagittal transverse* plane. (circle)
2. Forearm supination occurs in the *frontal sagittal transverse* plane. (circle)
3. Shoulder abduction occurs in the *frontal sagittal transverse* plane. (circle)
4. Doing jumping jacks causes movement of the arms and legs in the *frontal sagittal transverse* plane. (circle)
5. Doing sit-ups occurs in the *frontal sagittal transverse* plane. (circle)
6. Shaking your head 'no' produces movements in the *frontal sagittal transverse* plane. (circle)
7. Nodding your head 'yes' produces movements in the *frontal sagittal transverse* plane. (circle)
8. Shrugging the shoulders causes *elevation depression* of the scapula. (circle)
9. Clenching your jaw causes *elevation depression* of the mandible. (circle)

Label the following ligaments and joint structures of the knee, shoulder, elbow, and ankle.

Knee Joint: Label and color the following ligaments and structures.

Anterior cruciate ligament (ACL)

Medial collateral ligament (MCL)

Lateral collateral ligament (LCL)

Posterior cruciate ligament (PCL)

Medial meniscus

Lateral meniscus

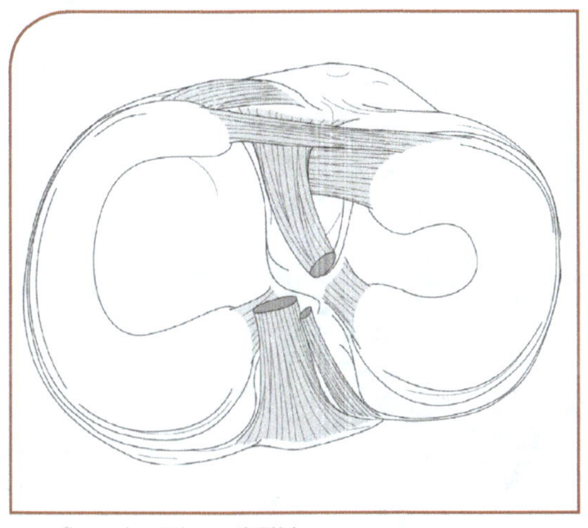

Superior View of Tibia

Illustration by Jamey Garbett. © 2003 Mark Nielsen

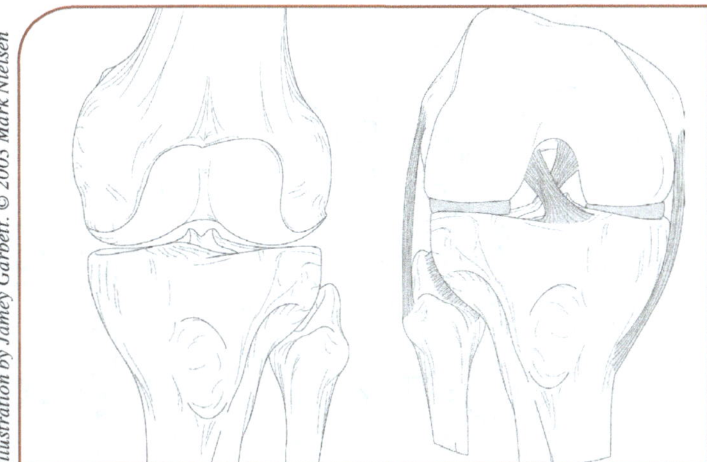

Posterior View, Anterior View

Talocrural Joint: Draw in the following ligaments using your textbook as a resource.

Medial or deltoid ligament

Anterior talofibular ligament

Calcaneofibular ligament

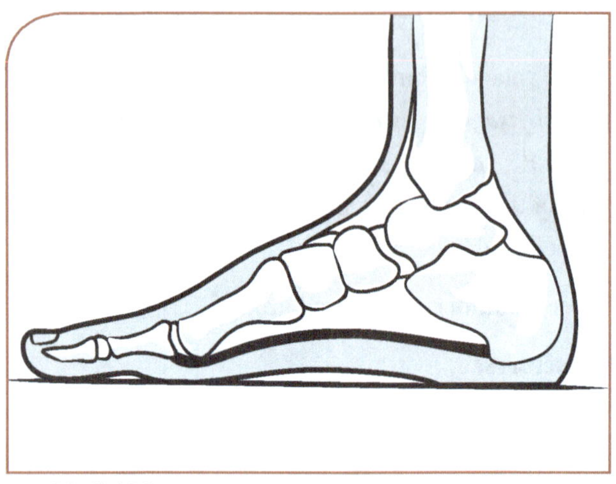

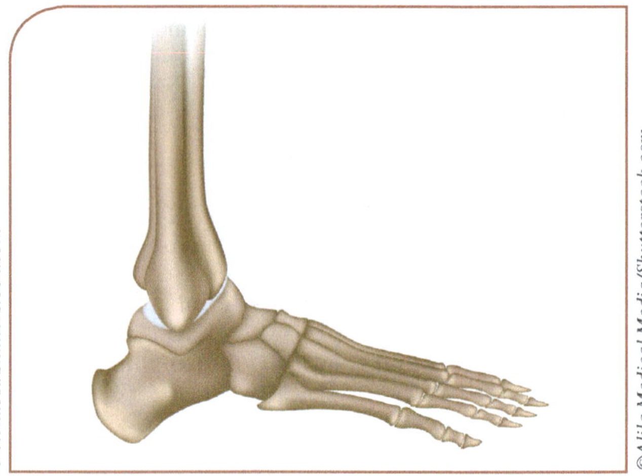

◾ Medial View   ◾ Lateral View

Hip Joint

Inguinal ligament

Labrum

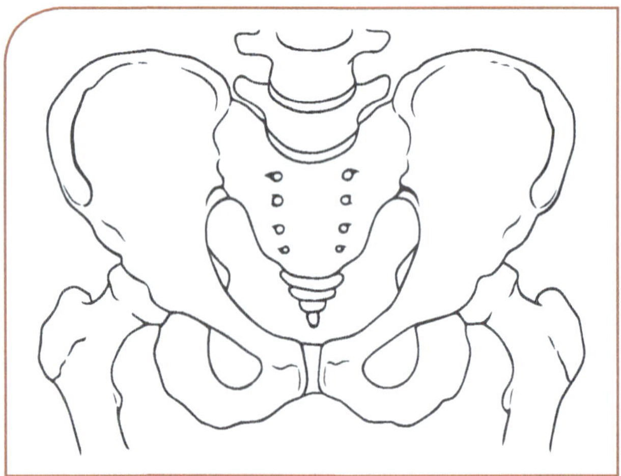

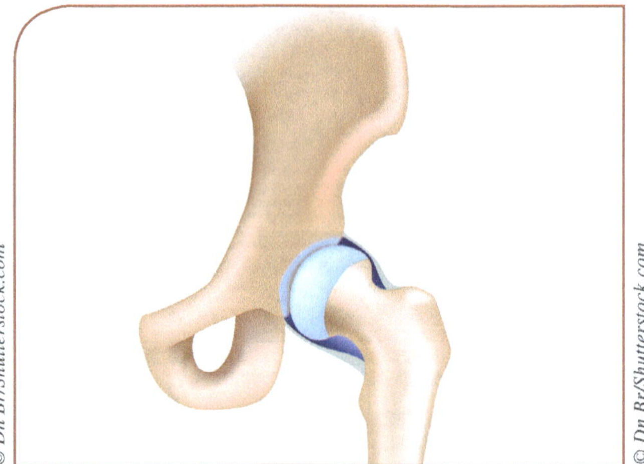

◾ Anterior View   ◾ Anterolateral View with Open Joint Capsule

Glenohumeral and Acromioclavicular Joint: Label the following ligaments and structures.

Acromioclavicular ligament

Coracoacrominal ligament

Coracoclavicular ligament

Deltoid bursa

Labrum

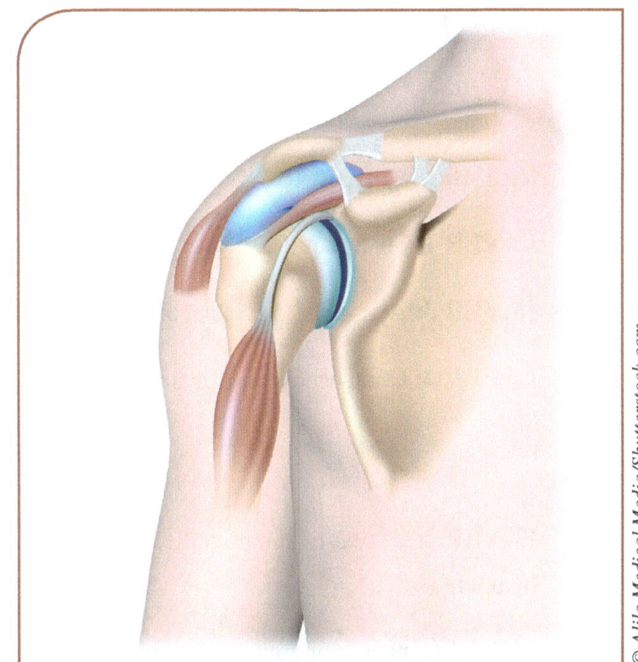

Anterior View

Humeroulnar and Humeroradial and Proximal Radioulnar Joints

Anular ligament

Medial collateral ligament (MCL)

Lateral collateral ligament (LCL)

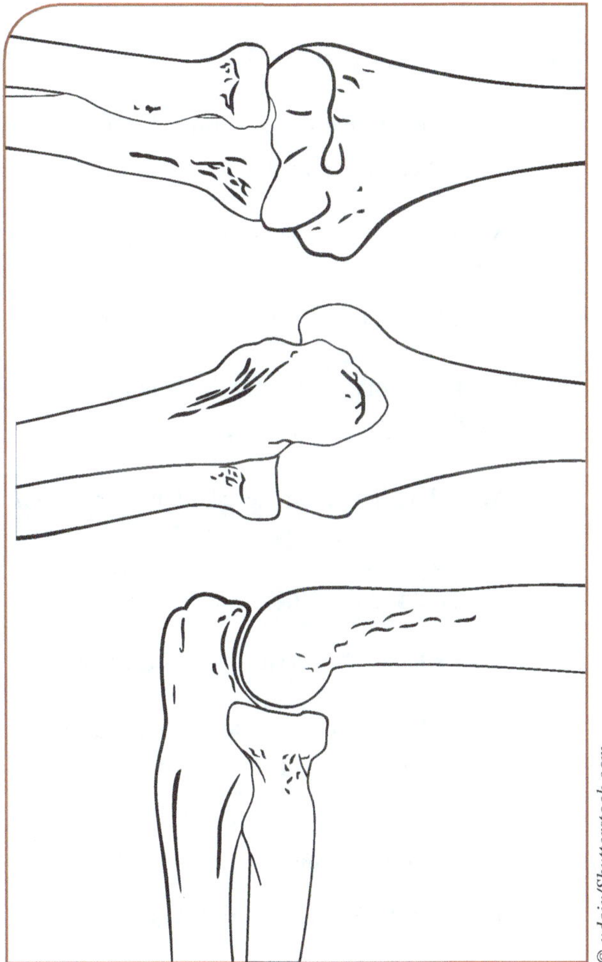

Anterior View, Posterior View, Lateral View

MODULE 5: SKELETAL

Dissection

Glenohumeral Joint

LAYER 1: Surface projection of acromion, surface projection of coracoid process

LAYER 3: Acromion, clavicle, coracoacromial ligament, coracoid

LAYER 4: Coracoclavicular ligament, fibrous capsule of glenohumeral joint

Elbow Joint

LAYER 4: Anular ligament, radial or lateral collateral ligament (LCL), ulnar or medial collateral ligament (MCL)

Which ligament maintains contact of the head of radius with the ulna? _______________________

Which ligament is often torn in baseball pitchers? ____

Knee Joint: Anterior

LAYER 4: Fibular or lateral collateral ligament (LCL), medial meniscus, lateral meniscus, tibial or medial collateral ligament (MCL)

What is the "unhappy triad"? ____

LAYER 5: Anterior cruciate ligament (ACL), posterior cruciate ligament (PCL)

Which ligaments prevent the femur from sliding forward and backward on the tibia?
ACL   PCL   MCL   LCL   (circle)

What ligament is stronger and thicker?   *ACL   PCL*   (circle)

Knee Joint: Posterior

LAYER 2: Fibrous capsule of the knee

LAYER 4: Fibular or lateral collateral ligament (LCL), medial meniscus, lateral meniscus, posterior cruciate ligament, tibial or medial collateral ligament (MCL)

Ankle Joint: Lateral

LAYER 2: Extensor retinaculum

LAYER 4: Lateral ligament of the ankle (the mostly injured ligaments are the anterior talofibular ligament and calcaneofibular ligament)

Ankle Joint: Medial

LAYER 4: Medial or deltoid ligament of the ankle

Histology: Hyaline Cartilage: Medium Magnification

Chondrocyte in lacuna, extracellular matrix, perichondrium

Animation: Synovial Joint

What is the most common type of joint? ____

List the six types of joints.

_______________________________   _______________________________

_______________________________   _______________________________

_______________________________   _______________________________

Four joint features in all synovial joints include:

_______________________________   _______________________________

_______________________________   _______________________________

A joint capsule has a fibrous layer that attaches to the _______________________________________.

The inside of the joint capsule is lined a _______________ which produces and absorbs

_______________.

The purpose of this fluid is to ____.

What type of connective tissue is articular cartilage? _______________________________________

The meniscus is found in the _______________ joint and acts as a _______________ .

Bursa _______________ between structures that cross a joint.

Animation: Intervertebral Joints

Intervertebral joints are formed by articulations between _______________ and between the superior and inferior articular processes.

Nodding your head forward is an example of cervical _______________________________________.

Looking up at the ceiling is an example of cervical _______________________________________.

"Side-ways" bending of the trunk is called _______________________________________.

Shaking you head no is an example of cervical _______________________________________.

Animation: Joint Movements—TMJ

The motion of opening the mouth is called _______________________________.

The motion of closing the mouth is called _______________________________.

Animation: Joint Movements—Scapula

The scapula has an articulation with the axial skeleton.   *TRUE   FALSE*   (circle)

An example of scapular elevation is _______________________________.

The opposite of scapular elevation is scapular _______________________________.

Bringing the border of the scapulae closer together is an example of scapular _______________________.

The opposite of scapular retraction is scapular _______________________________.

Animation: Joint Movements—Glenohumeral Joint

The glenohumeral (GH) joint is the articulation between the _______________ and the

_______________ of the humerus.

Swinging the arm forward is an example of GH _______________________________.

Glenohumeral abduction is moving the arm away from the body in the _______________ plane.

The opposite of glenohumeral abduction is glenohumeral _______________________________.

Medial or internal rotation is turning the humerus _______________ the body compared to lateral or

external rotation which is turning the humerus _______________ from the body.

What is a compound motion of the shoulder that results in a movement in a circular pattern?

____

Animation: Joint Movements—Elbow

The elbow joint is formed by two articulations. These articulations are:

Between the _______________ of the ulna and the _______________ of the humerus

Between the _______________ of the radius and the _______________ of the humerus.

The elbow joint is a hinge joint and can only move in _______________ and _______________.

Animation: Joint Movements—Radioulnar

The radioulnar joint is an articulation between the ________________ and ________________ radius and ulna.

When the palm faces anteriorly, the motion is called ____.

When the palm faces posteriorly, the motion is called ____.

The radius and ulna are parallel in   *supination   pronation*   . (circle)

Animation: Joint Movements—Radiocarpal (Wrist)

The ulna articulates with the carpal bones.   *TRUE   FALSE*   (circle)

When curling a hand weight would be an example of wrist ________________________________.

When moving the wrist sideways to the pinky side, it is called ________________________________.

When moving the wrist sideways to the thumb side, it is called ________________________________.

Animation: Joint Movements—CMC of Thumb

The CMC joint of the thumb is between a carpal bone and the base of the ________________________.

This unique saddle joint allows for ________________ to occur which is movement of the thumb to the pinky.

Animation: Joint Movements—MCP of Fingers

The MCP joint is an articulation between the heads of the ________________ and the

________________.

Making a fist would cause the MCP to move into ____.

Spreading the fingers out would cause the MCP to move into ________________________________.

Animation: Joint Movements—IP of Fingers

The IP joints are articulations between the phalanges. The PIPs (proximal interphalangeal joint) is

between the ________________ and the ________________ phalanges. The DIPs (distal

interphalangeal joint) is between the ________________ and the ________________ phalanges.

These joints are capable of moving into ________________ and ________________.

Animation: Joint Movements—Hip

The hip joint is an articulation between the ________________ of the femur and the

________________.

Movement of the hip in a posterior direction is called hip ________________________________.

Movement of the hip in a sideways direction is called hip ________________________________.

________________ of the hip is a compound circular motion.

Rotating the hip in a ________________ rotation will point the patella toward the inside of the leg.

Animation: Joint Movements—Knee

Does the fibula participate in the movement in the knee joint?   *YES   NO*   (circle)

The motion that describes the knee bending is called knee ________________________.

The motion that describes the knee straightening is called knee ________________________.

Animation: Joint Movements—Tibiofibulotalar (Ankle)

This joint is also called the talocrural joint.

Moving the dorsum of the foot toward the shin is called ________________________.

Moving the plantar surface of the foot away from the shin is called ________________________.

Animation: Joint Movements—Intertarsal

Moving the plantar surface of the foot inward is called ________________________.

Moving the plantar surface of the foot outward is called ________________________.

IN LAB ACTIVITIES

1. Discussion on subluxation vs. dislocation

   Video on Glenohumeral Joint Dislocation

   Subluxations vs. Dislocations

   Subluxation: A joint injury in which there is partial contact between the two surfaces of the bones. You may consider this to be a partial dislocation.

   Dislocation: A joint injury in which the articulations have no contact with each other.

   Reduction: Surgical or manipulative repositioning of bones into anatomical alignment.

2. Label whether the lower extremities are in genu varus, genu valgus, or normal.

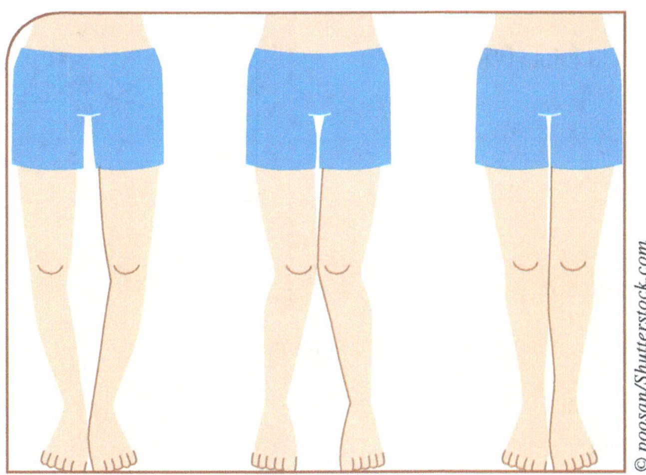

© poosan/Shutterstock.com

3. **Stations and Check-offs**

Using your Pre-lab as a guide. Build and/or identify the following structures at each station.

_________________ Station 1:

On mini skeleton, identify the following joints. When tested you cannot refer to your sheet.

1. Acromioclavicular joint (AC joint)
2. Carpometacarpal joints (CMC)
3. Cervical facet joints (i.e. C5/C6)
4. Cervical intervertebral joints
5. Distal interphalangeal joint (DIP)
6. Glenohumeral (GH joint)
7. Humeroulnar and humeroradial (elbow joint)
8. Intercarpal joints
9. Metacarpophalangeal (MCP)
10. Proximal interphalangeal joint (PIP)
11. Radioulnar joints (proximal and distal)
12. Scapulothoracic (functional joint—not a real synovial joint)
13. Sternoclavicular joint (SC joint)
14. Temporomandibular joint (TMJ)
15. Wrist joint (radiocarpal joint)

_________________ Station 2:

On the mini skeleton, identify the following joints. When tested you cannot refer to your sheet.

1. Distal interphalangeal joint (DIP)
2. Hip (coxal)
3. Knee joint
4. Lumbar facet joints (i.e. L2/L3)
5. Lumbar intervertebral joints

6. Metatarsophalangeal joint (MTP)
7. Patellofemoral joint
8. Proximal interphalangeal joint (PIP)
9. Pubic symphysis
10. Sacroiliac joint (SI joint)
11. Talocrural and Subtalar joints
12. Thoracic facet joints

_________________ Station 3:

Build the acromioclavicular ligaments and elbow ligaments

1. Acromioclavicular ligaments
2. Coracoacromial ligament
3. Coracoclavicular ligament
4. Acromioclavicular ligament
5. Glenohumeral labrum (using tacky dough)
6. Elbow Ligaments
7. Anular ligament
8. MCL
9. LCL

_________________ Station 4:

Build the hip, knee, and talocrural ligaments

Hip
Inguinal ligament
Labrum (using tacky dough)

Knee
Medial meniscus
Lateral meniscus
ACL
PCL
MCL
LCL

Ankle
Calcaneofibular ligament (on the medial side)
Anterior talofibular ligament (on the medial side)
Deltoid ligament (on lateral side)

_________________ Station 5:

Watch the videos and perform the following tests. This should be done with a partner.

TESTS OF THE KNEE

Lachmans Test: Test Integrity of ACL

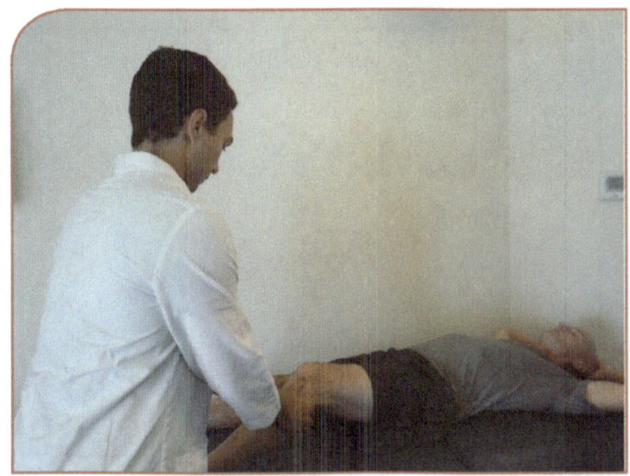

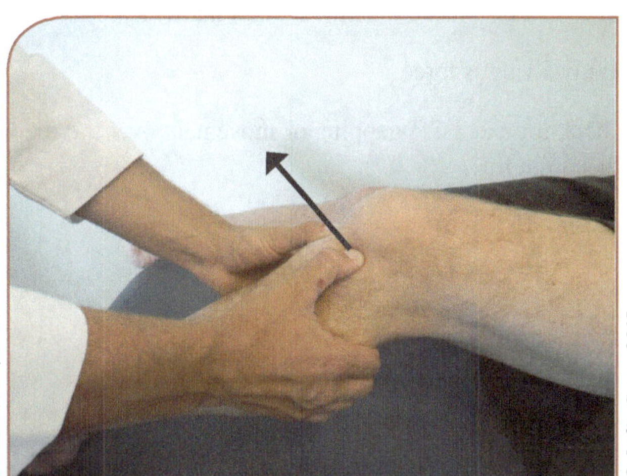

Position patient supine with the knee at 30 degree angle.

Apply anterior force to the proximal tibia.

Watch for anterior movement of the tibia on the femur.

(+) for ACL tear

Posterior Drawer Test: Test Integrity of PCL

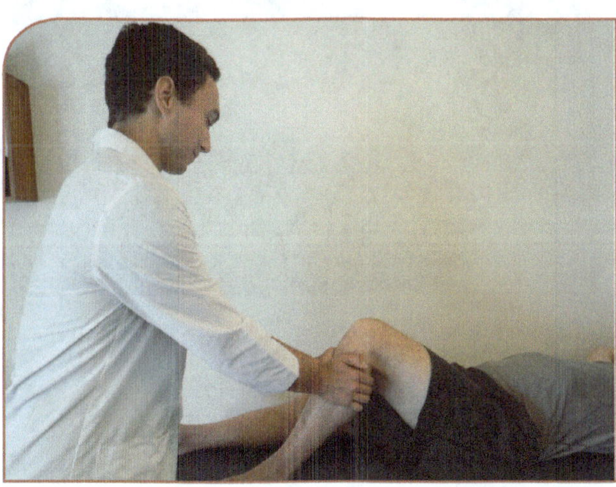

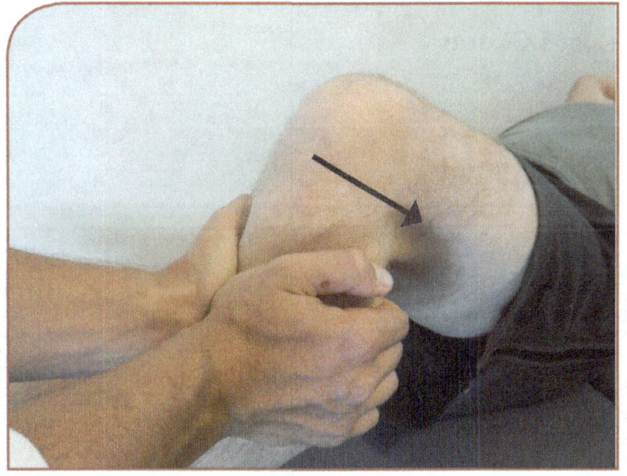

Position patient supine with knee at 90 degree angle and foot flat on table.

Apply posterior force on tibial tuberosity.

Watch for posterior movement of tibia on the femur.

(+) for PCL tear

Valgus Stress Test: Test Integrity of MCL

Position patient supine with knee slightly flexed.

One hand is placed over the lateral knee joint line and the other hand is placed over the medial malleolus.

Apply a valgus force.

Check for excessive gapping or movement over the medial knee joint line.

(+) for MCL tear

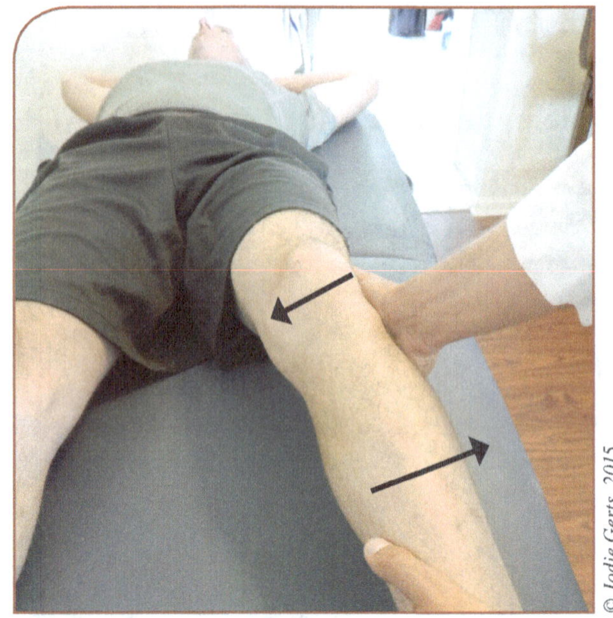

© Jodie Gerts, 2015

Varus Stress Test: Test Integrity of LCL

Position patient supine with knee slightly flexed.

One hand is placed over the medial knee joint line and the other hand is placed over the lateral malleolus.

Apply a varus force.

Check for excessive gapping or movement over the lateral knee joint line.

(+) for LCL tear

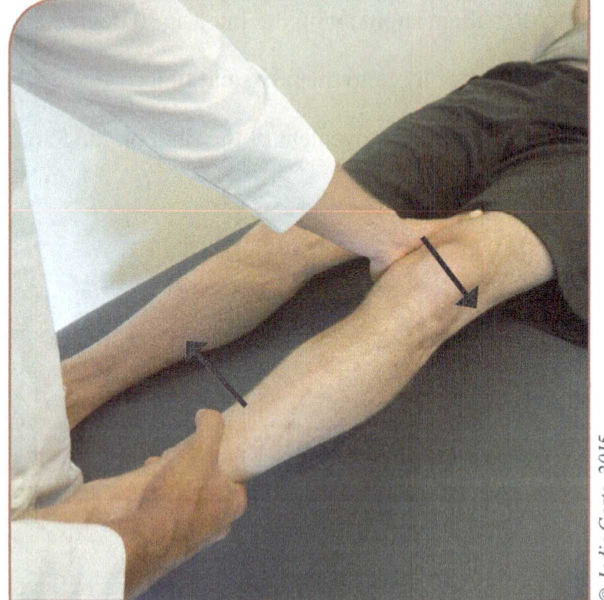

© Jodie Gerts, 2015

TESTS OF THE ANKLE

Anterior Drawer Test: Test Integrity of Anterior Talofibular Ligament and/or Calcaneofibular Ligament.

Position patient supine with ankle in neutral (90 degrees).

Grasp the calcaneous with one hand and grasp the anterior medial and lateral malleoli with the other hand.

Apply anterior force to the calcaneous.

Check for excessive motion of the talus sliding anterior on the tibia and fibula.

(+) for anterior talofibular ligament tear

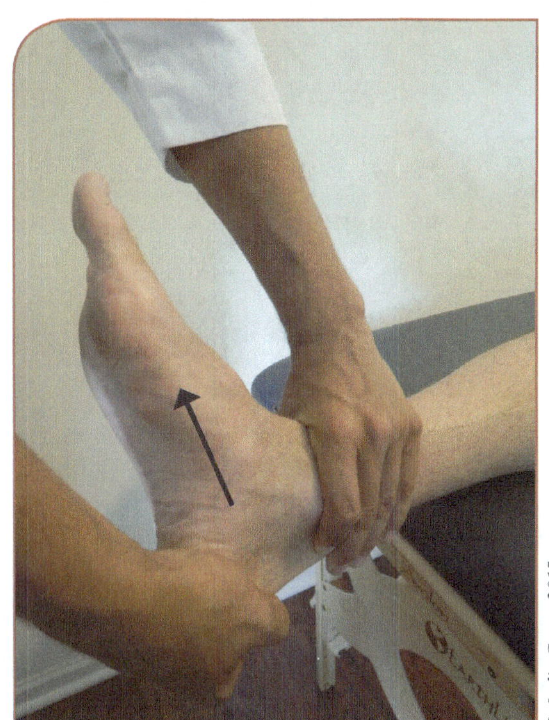

L A B *7*

The Muscular System—
Head, Neck and Lower Extremity

Objectives

The purpose of this lab is to have the student know the anatomy and function of the facial, trunk and lower extremity muscles. The student will also identify microanatomical structures of the muscle. The student will be able to integrate the skeletal information from Lab 5 to identify basic fracture types from radiographs and different fixation treatments.

Prior to lab: Complete the Pre-lab.

Complete APR Dissection, Histology and Animations

During lab: Complete each objective prior to leaving lab.

- [ ] **Sliding Filament Model**
- [ ] **Build the Leg Muscles on the Skeleton**
- [ ] **Radiographs**
    Classify the Fracture
    State the Type of Treatment

LAB 7 PRELAB—The Muscular System—Head, Neck and Lower Extremity

This is a table of the origins and insertions for the lower extremity muscles. Refer to this table when completing the pre-lab.

Origins and Insertions for Lower Extremity Muscles

Muscle	Action	Origin	Insertion
Iliopsoas	Flexion of hip	Iliac crest and bodies of lumbar vertebrae	Lesser trochanter
Adductor muscle. group	Adductor of hip	Pubis	Medial femoral shaft
Gluteus maximus	Extension of hip	PSIS and posterior sacrum	Iliotibial band (and greater trochanter)
Gluteus medius	Abduction of hip	Iliac crest lateral	Greater trochanter
Piriformis	External rotation of hip	Sacrum	Greater trochanter
Tensor fascia latae	Flexion and abduction of hip	Iliac crest and ASIS	Iliotibial band (greater trochanter)
Sartorius	Flexion and abduction of hip	ASIS	Medial, proximal tibia
Quadriceps group	Extension of the knee	AIIS	Tibial tuberosity
Tibialis anterior	Dorsiflexion of ankle	Proximal anterior lateral shaft of tibia	Base of 1st metatarsal
Fibularis longus	Eversion of ankle	Proximal, lateral surface of fibula	Base of 1st metatarsal
Extensors of toes	Extension of toes	Proximal, lateral tibia and fibula	Plantar distal phalanx of toes
Hamstring group	Flexion of knee	Ischial tuberosity	Fibular head Medial tibial condyle
Gastrocnemius	Flexion of fingers	Femoral epicondyles	Posterior calcaneous
Soleus	Extension of fingers	Proximal, posterior tibia	Posterior calcaneous
Flexors of toes	Flexion of the toes	Mid, posterior fibula	Dorsum of distal phalanx of toes
Intinsic muscles of foot	Finger movement	Bones of the foot	Bones of the foot
Iliotibial band	Extension of trunk	Iliac crest (with gluteus maximus and TFL)	Lateral proximal tibia

Define the origin of a muscle: ____

Define the insertion of a muscle: ____

Define the action of a muscle: ____

Label these muscles on the picture. Color the origin in RED and color the insertion in BLUE.

Iliopsoas

Adductor muscles (adductor longus, adductor magnus, adductor brevis). You do not need to know the individual muscles.

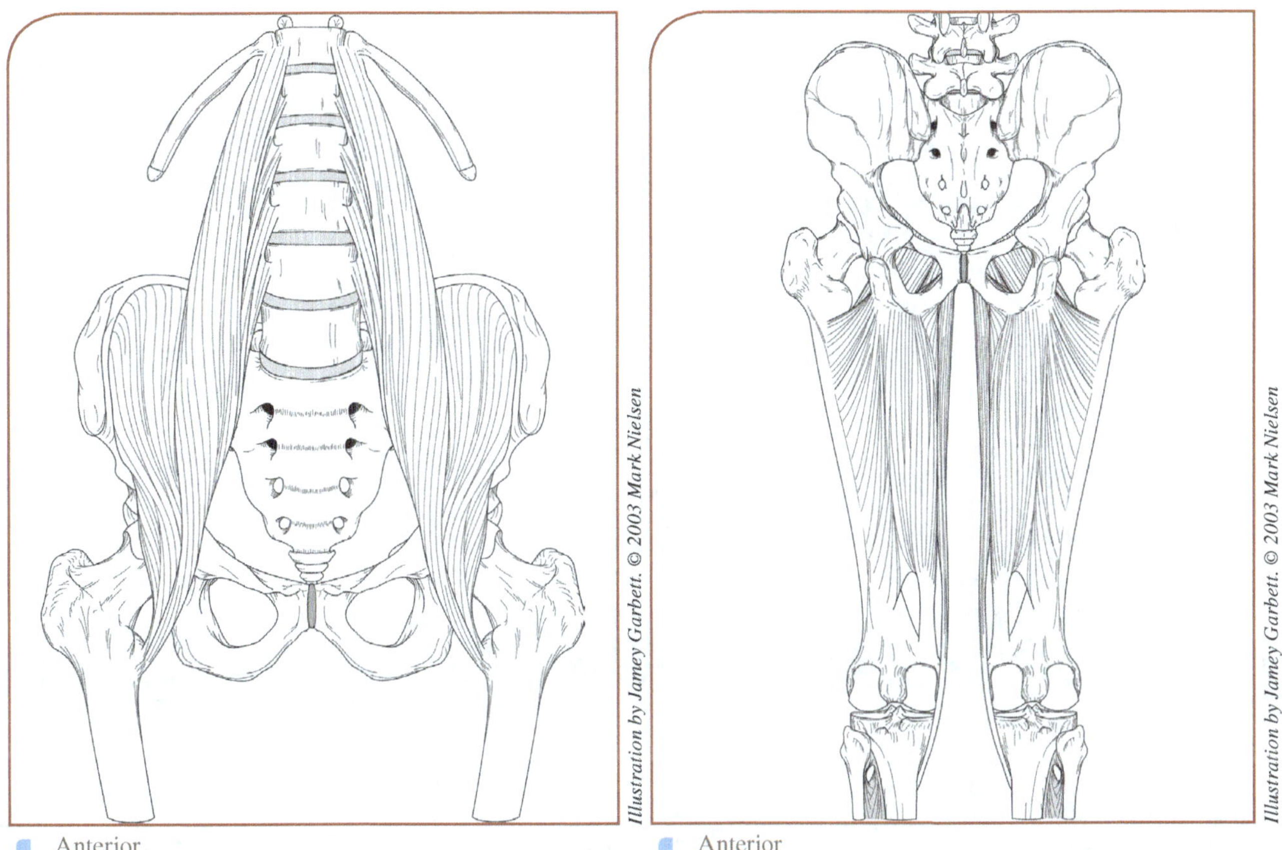

Illustration by Jamey Garbett. © 2003 Mark Nielsen

Anterior

Illustration by Jamey Garbett. © 2003 Mark Nielsen

Anterior

State the action and the joint moved of the following muscles:

1. Iliopsoas: flexion of the hip

2. Adductor muscles: ____

Label these muscles and other structures on the picture. Color the origin in RED and color the insertion in BLUE.

Gluteus maximus

Gluteus medius

Hamstrings group (biceps femoris, semitendinosus, semimembranosus)

Piriformis

Tensor fascia latae

Iliotibial band (ITB)

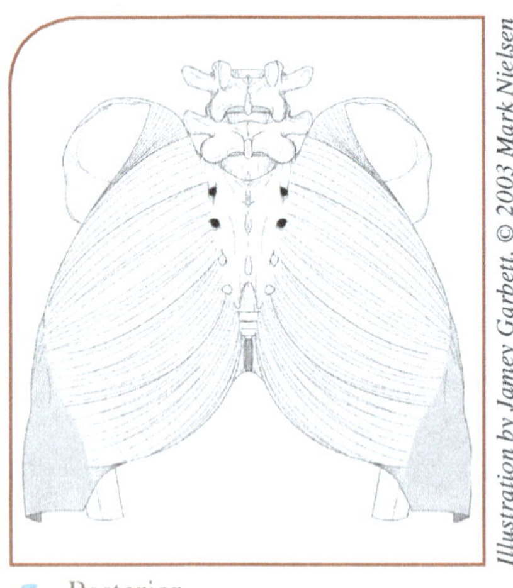

Posterior

Illustration by Jamey Garbett. © 2003 Mark Nielsen

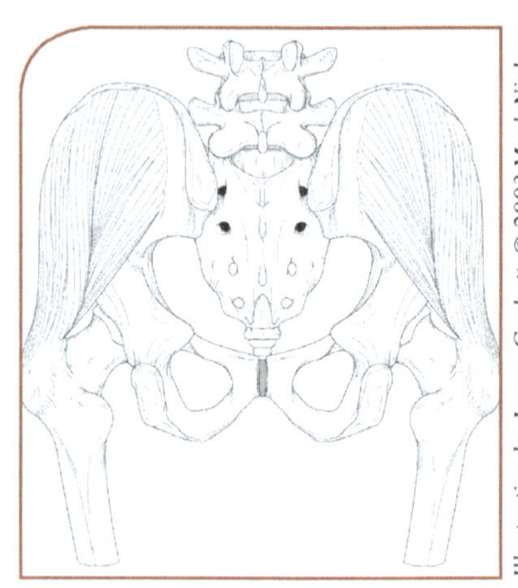

Posterior

Illustration by Jamey Garbett. © 2003 Mark Nielsen

Posterior

Illustration by Jamey Garbett. © 2003 Mark Nielsen

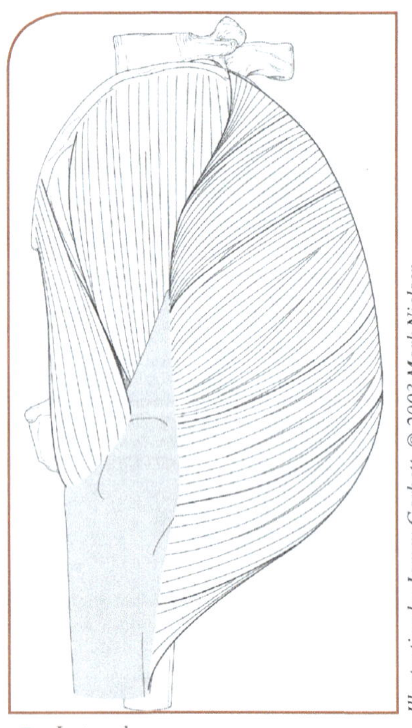

Lateral

Illustration by Jamey Garbett. © 2003 Mark Nielsen

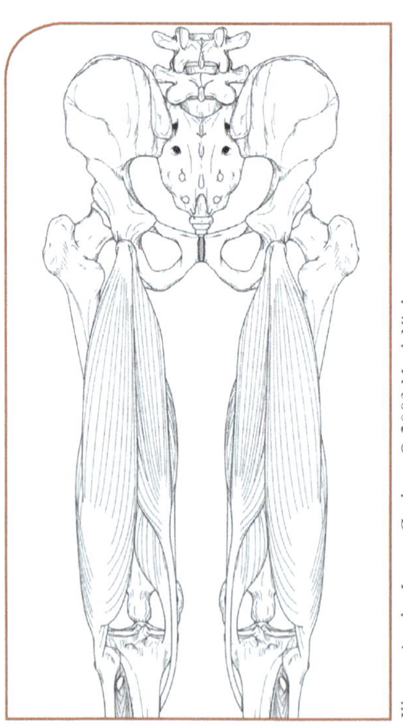

Posterior

Illustration by Jamey Garbett. © 2003 Mark Nielsen

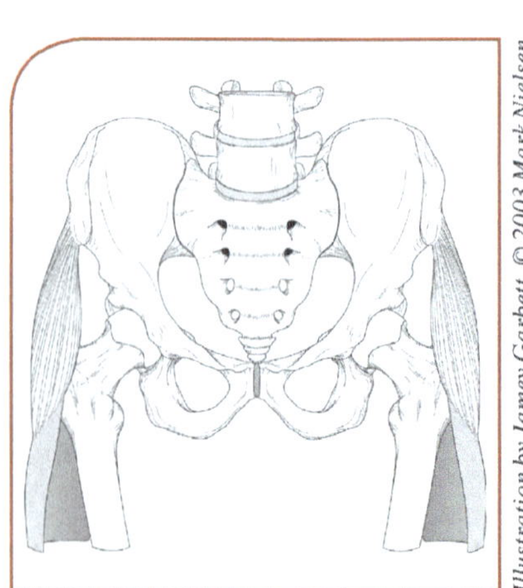

Anterior

Illustration by Jamey Garbett. © 2003 Mark Nielsen

State the action and the joint moved of the following muscles:

1. Gluteus maximus: ____

2. Gluteus medius: ____

3. Hamstring group: ____

4. TFL: ____

Label these muscles and other structures on the picture. Color the origin in RED and color the insertion in BLUE.

Quadriceps group

Sartorius

Patellar tendon or patellar ligament

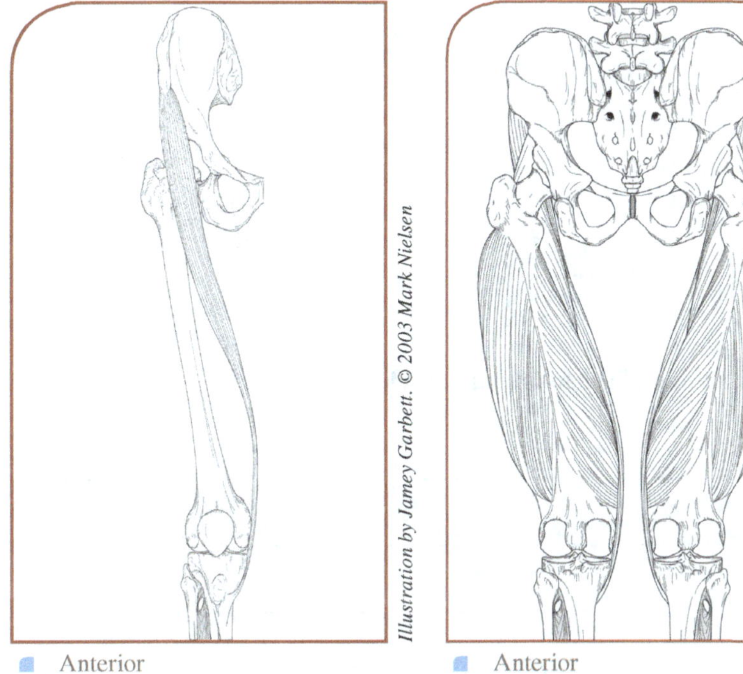

State the action and the joint moved of the following muscles:

1. Quadriceps group: ____

2. Sartorius: ____

Label these muscles and other structures on the picture. Color the origin in RED and color the insertion in BLUE.

Tibialis anterior

Fibularis longus

Extensors of the toes (extensor digitorum longus)

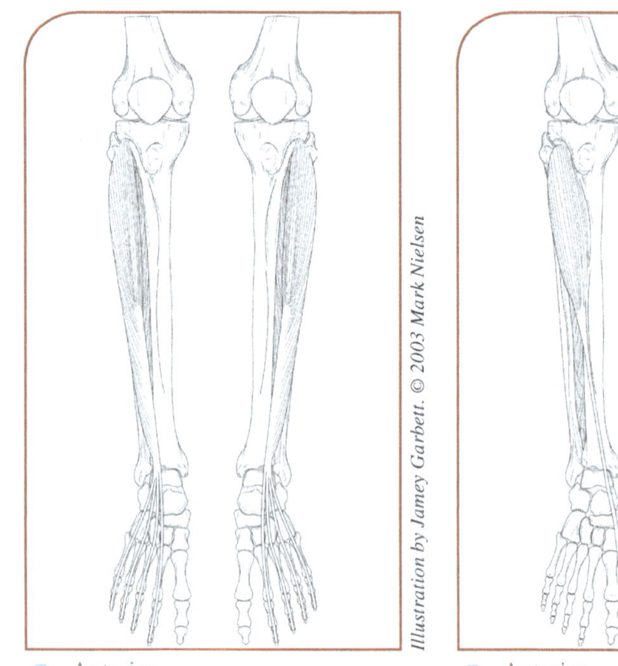

Anterior

Illustration by Jamey Garbett. © 2003 Mark Nielsen

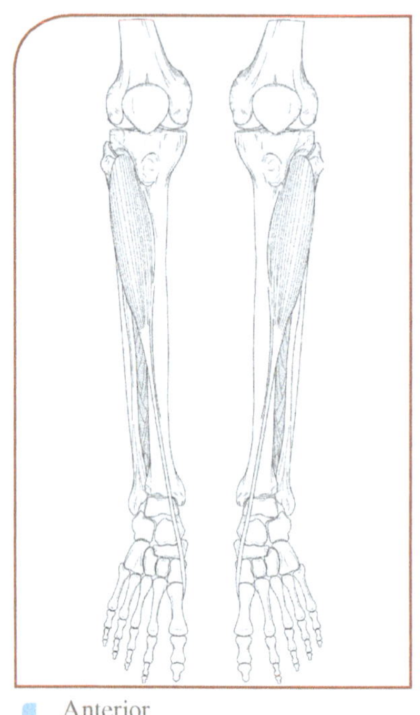

Anterior

Illustration by Jamey Garbett. © 2003 Mark Nielsen

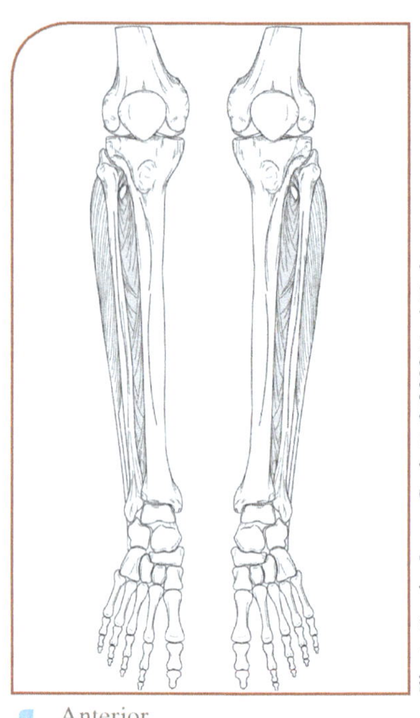

Anterior

Illustration by Jamey Garbett. © 2003 Mark Nielsen

State the action and the joint moved of the following muscles:

1. Tibialis anterior: ____

2. Fibularis longus: ____

3. Extensors of the toes: ____

Label these muscles and other structures on the picture. Color the origin in RED and color the insertion in BLUE.

Gastrocnemius

Soleus

Calcaneal tendon (Achilles tendon)

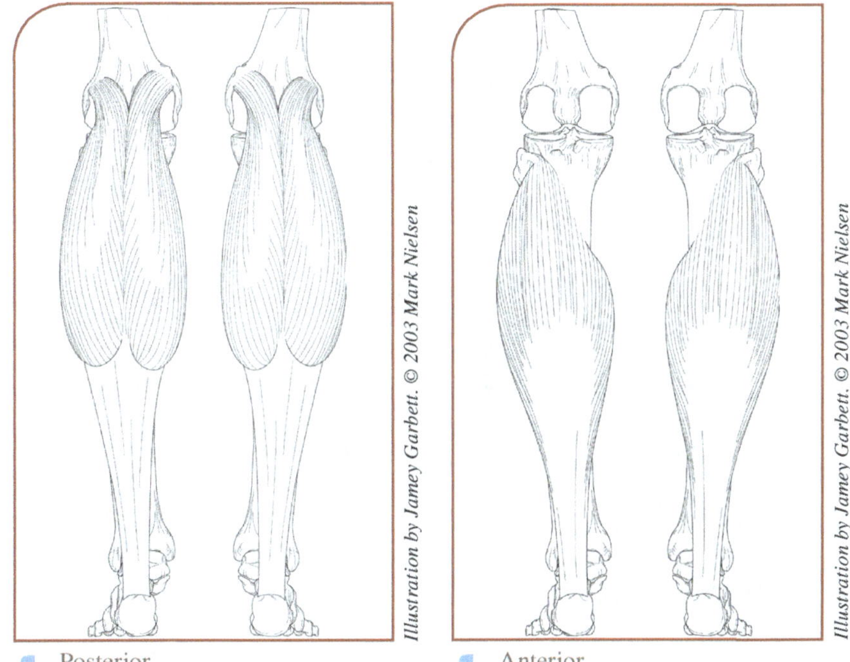

Posterior

Anterior

Illustration by Jamey Garbett. © 2003 Mark Nielsen

State the action and the joint moved of the following muscles:

1. Gastrocnemius: plantarflexion of the talocrural joint (ankle joint)

2. Soleus: ____

Label these muscles and other structures on the picture. Color the origin in RED and color the insertion in BLUE.

Flexors of the toes

Intrinsic muscles of the foot

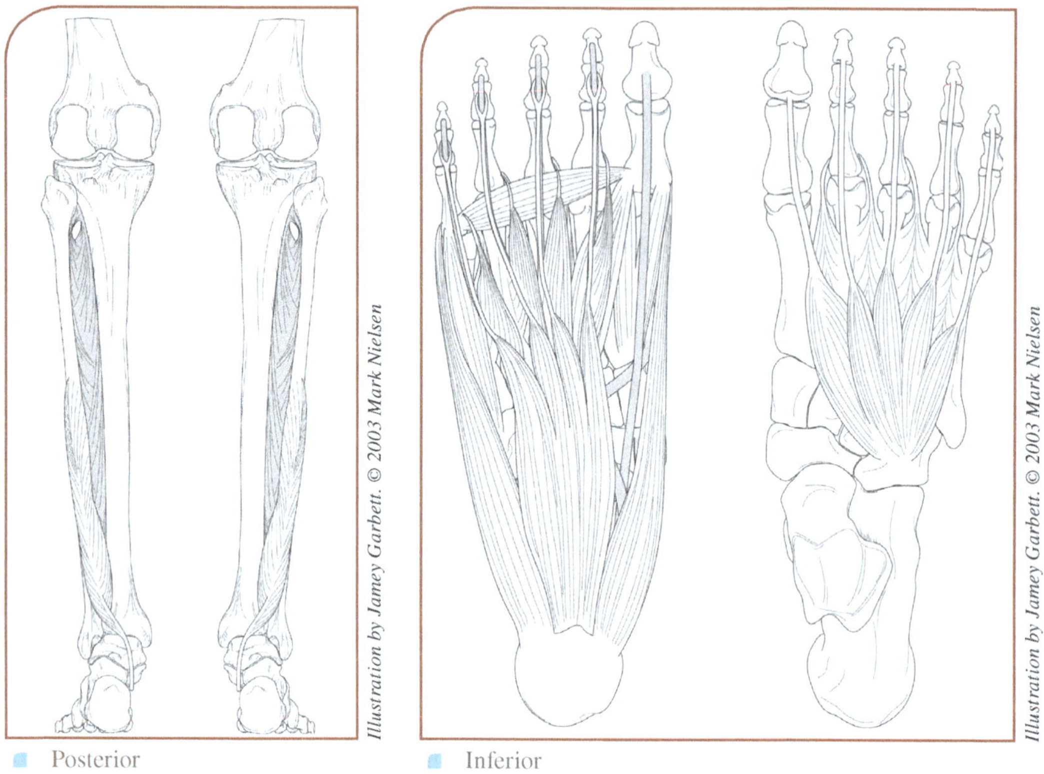

State the action and the joint moved of the following muscles:

Flexors of the toes: ____

These pictures are the superficial muscles of the anterior and posterior lower extremity muscles. Once you have completed coloring the individual muscles, color, and label these pictures.

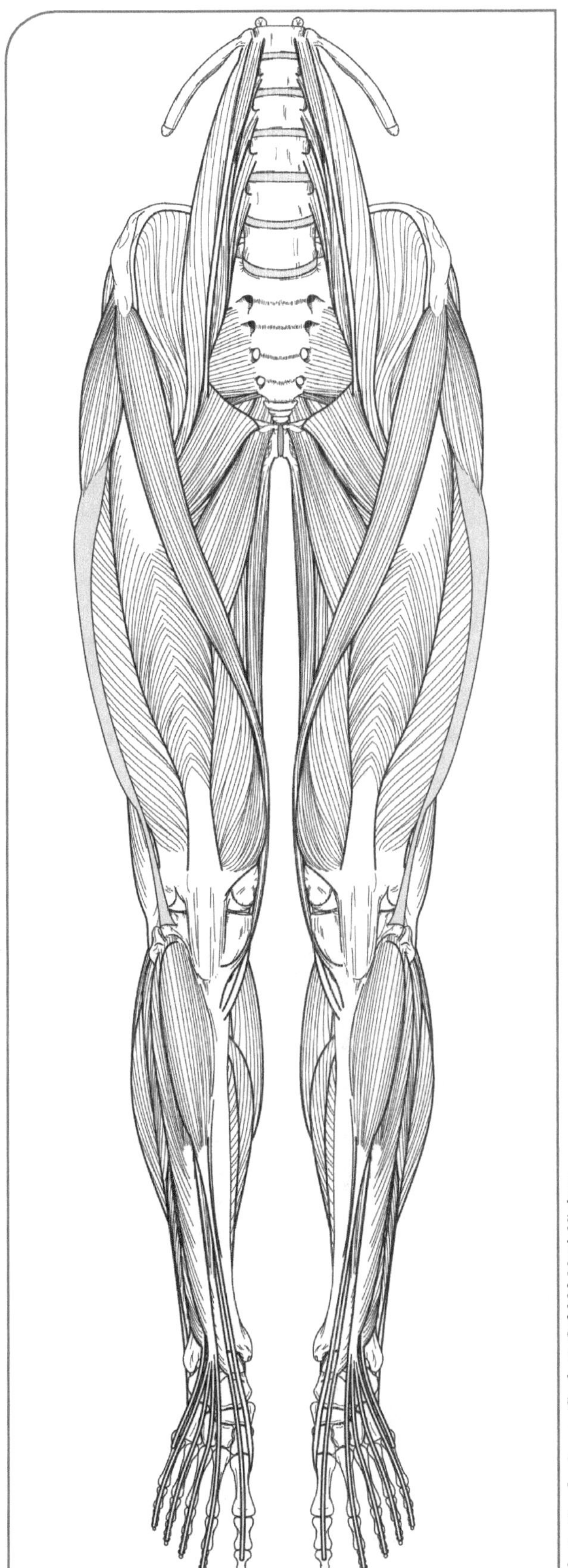

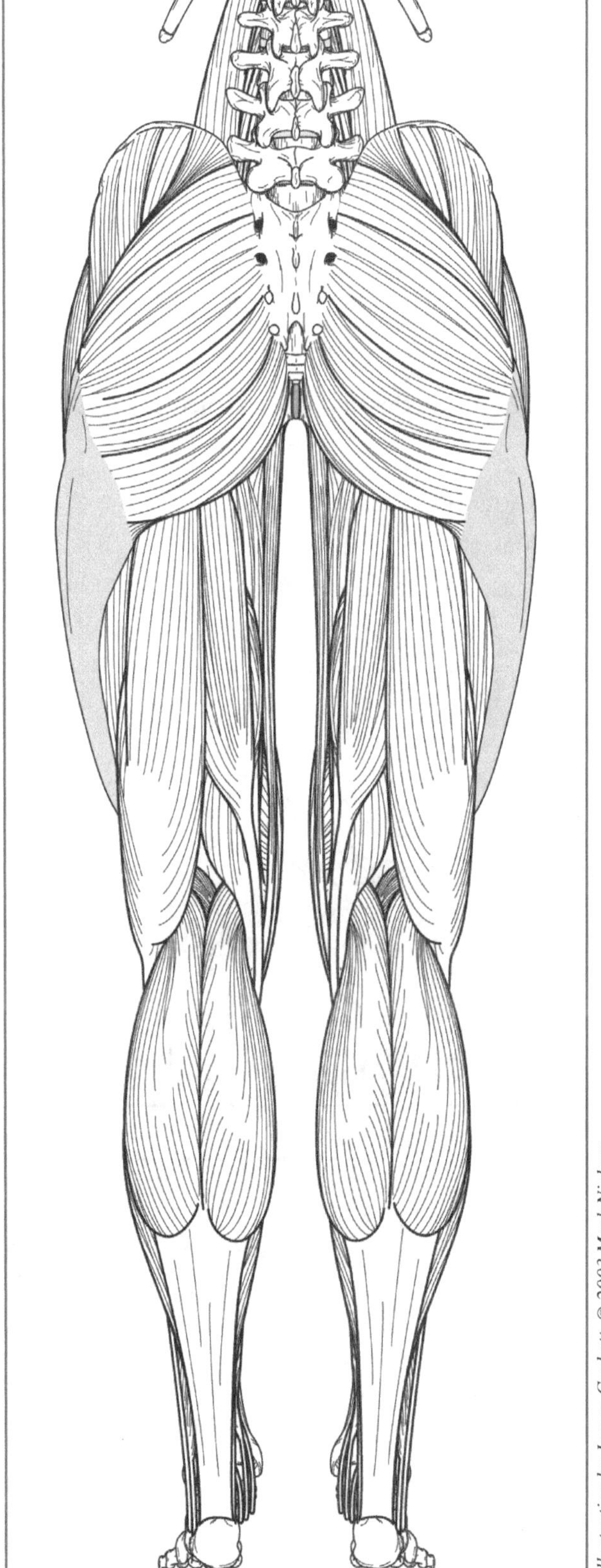

This is a table of the origins and insertions for the head and neck muscles. Refer to this table when completing the pre-lab.

Muscle	Action	Origin	Insertion
Extraocular muscles	Movement of the eyeball	Orbit of eye	Eye
Muscles of facial expression	Movement of the face	Facial bones	Dermis of face
Masseter	Elevation of mandible	Zygomatic arch	Inferior mandible
Temporalis	Elevation of mandible	Temporal bone	Superior mandible
Sternocleidomastoid	Flexion of neck	Manubrium and clavicle	Mastoid process of temporal bone

Label these muscles and other structures on the picture. Color the origin in RED and color the insertion in BLUE.

Extraocular muscles

Masseter

Muscles of facial expression

Sternocleidomastoid

Temporalis

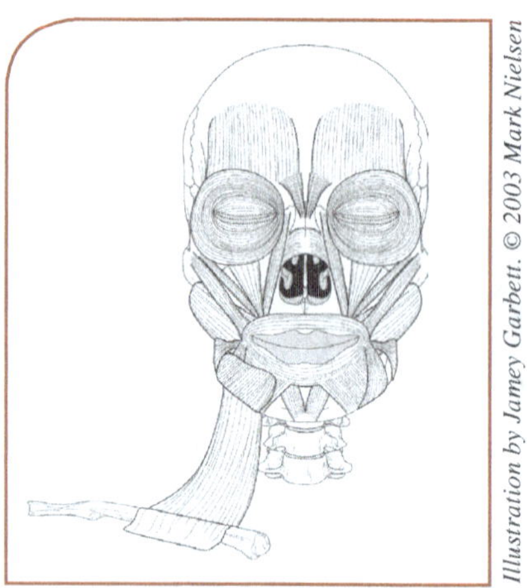

Illustration by Jamey Garbett. © 2003 Mark Nielsen

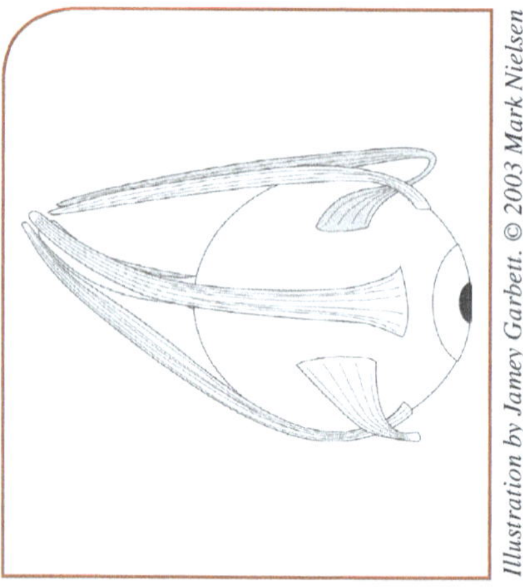

Illustration by Jamey Garbett. © 2003 Mark Nielsen

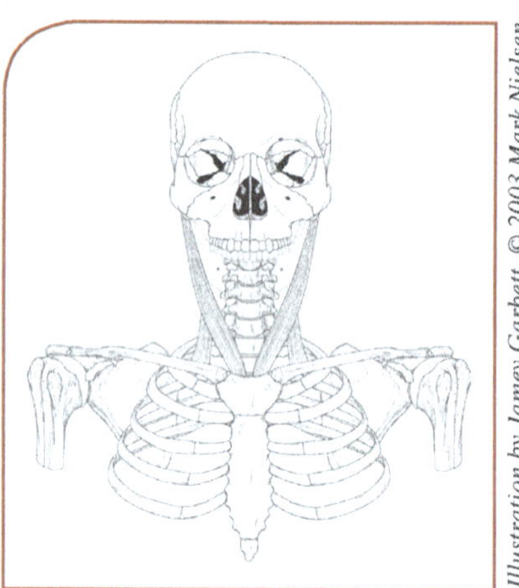

Illustration by Jamey Garbett. © 2003 Mark Nielsen

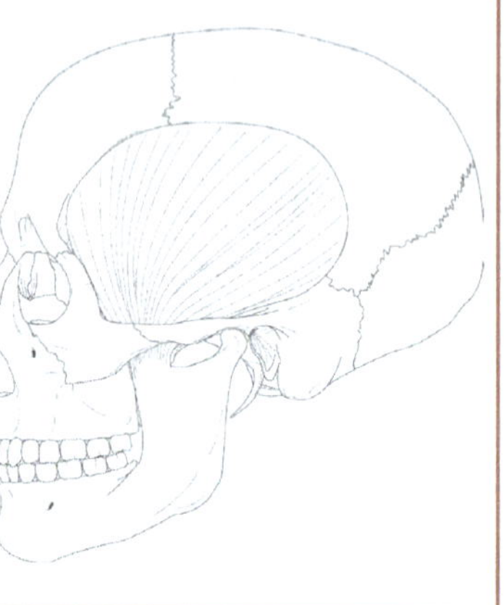

Illustration by Jamey Garbett. © 2003 Mark Nielsen

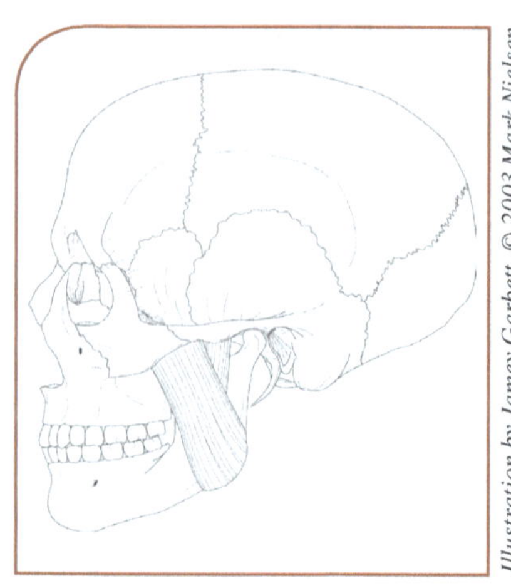

Illustration by Jamey Garbett. © 2003 Mark Nielsen

Label the following muscle microanatomy structures:

Thin filament

Thick filament

Z disc

Sarcomere

A band

I band

H band

M line

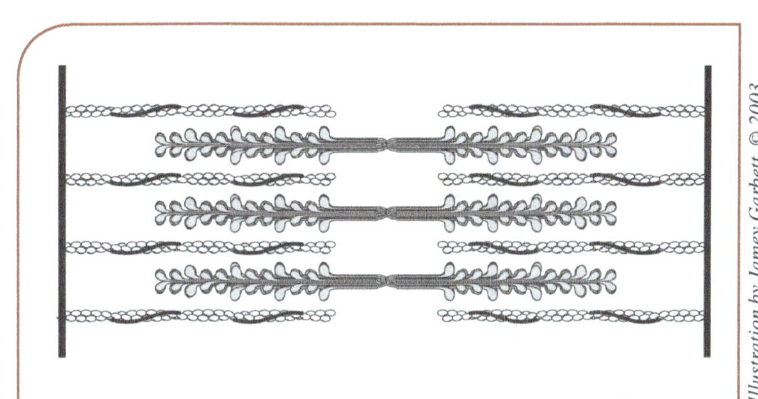

Label the following muscle microanatomy structures:

Thin filament:

  Actin

  Tropomyosin

  Troponin

Thick filament:

  Myosin molecule

  Myosin head

  Myosin tail

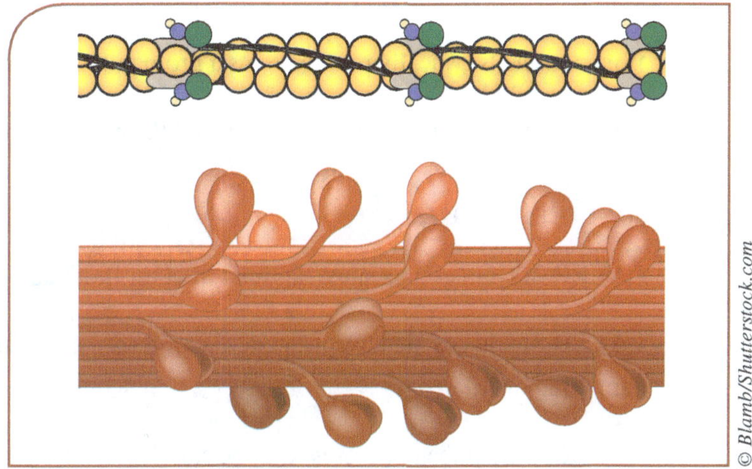

MODULE 6: THE MUSCULAR SYSTEM

Dissection

Head and Neck: Anterior

LAYER 2: All the muscles are muscles of facial expression

What do the muscles of facial expression have in common? ___________________________________

LAYER 3: Sternocleidomastoid

Head and Neck: Lateral

LAYER 1: Surface projection of sternocleidomastoid

LAYER 2: Masseter

LAYER 3: Sternocleidomastoid, temporalis

LAYER 4: All muscles in the orbit are extraocular muscles (EOM)

What is the function (action) of the extraocular muscles? ___________________________________

Abdomen

LAYER 5: Iliacus and psoas major can be called the iliopsoas muscle

Pelvis

LAYER 1: Iliacus and psoas major can be called the iliopsoas; pelvic floor muscles.

Hip and Thigh: Anterior

LAYER 2: Iliopsoas, quadriceps femoris, sartorius, tensor fascia latae

LAYER 3: Adductor longus can be called the adductor muscle of the hip

What muscle is typically referred to as a groin pull or groin strain? _______________________

LAYER 4: Adductor magnus can be called the adductor muscle of the hip

Hip and Thigh: Posterior

LAYER 2: Gluteus maximus, iliotibial tract (or iliotibial band)

LAYER 3: Gluteus medius, piriformis, semitendinosus and biceps femoris (can be called the hamstring muscle)

LAYER 4: Semimembranosus (Can be called the hamstring muscle)

How many muscles comprise the muscle group known as the hamstrings? _______________________

Leg and Foot: Anterior

LAYER 2: Tibialis anterior, fibularis longus, extensor digitorum (can be called the extensors of toes)

Leg and Foot: Posterior

LAYER 1: Calcaneal tendon

LAYER 2: Gastrocnemius, calcaneal tendon

What muscles make-up the calcaneal tendon? _______________________

What tendon is the strongest tendon in the body? _______________________

LAYER 3: Soleus

LAYER 4: Flexor digitorum longus (Can be called the flexors of toes)

Foot

LAYER 2: Plantar aponeurosis or plantar fascia

LAYER 3: All muscles in this slide are intrinsic muscles of foot

Histology: Skeletal Muscle: High Magnification

A band, I band, nucleus

Histology: Sarcomere

A band, H band, I band, M line, sarcomere, Z disc

Animation: Skeletal Muscle

Skeletal muscle is responsible for ___________________ movement.

Which connective tissue surrounds the whole muscle? ___________________________________

Perimyosium surrounds ____.

Muscle fibers are wrapped with a connective tissue called _______________________________

Myofibrils are made up of two kinds of protein filaments called ___________________ and

___________________ which are known are ___________________.

The contractile unit of muscle is called the ____.

The actin slides along the ___________________ which shortens the sarcomere causing a muscle

___________________.

Sliding Filament Model

Thin filament:
   Actin
   Tropomyosin
   Troponin
Thick filament:
   Myosin molecule
   Myosin head
Z disc
Sarcomere
A band
I band
H band
M line

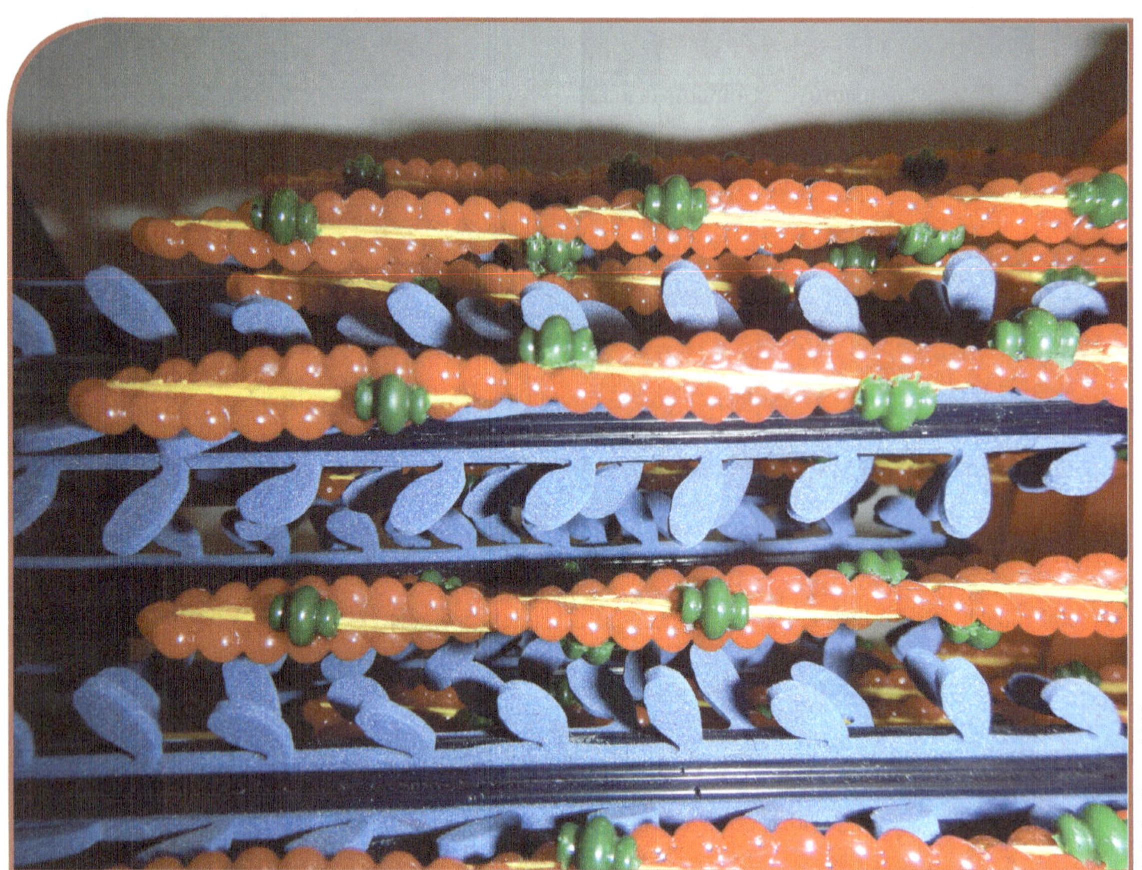

© Jodie Gerts, 2015

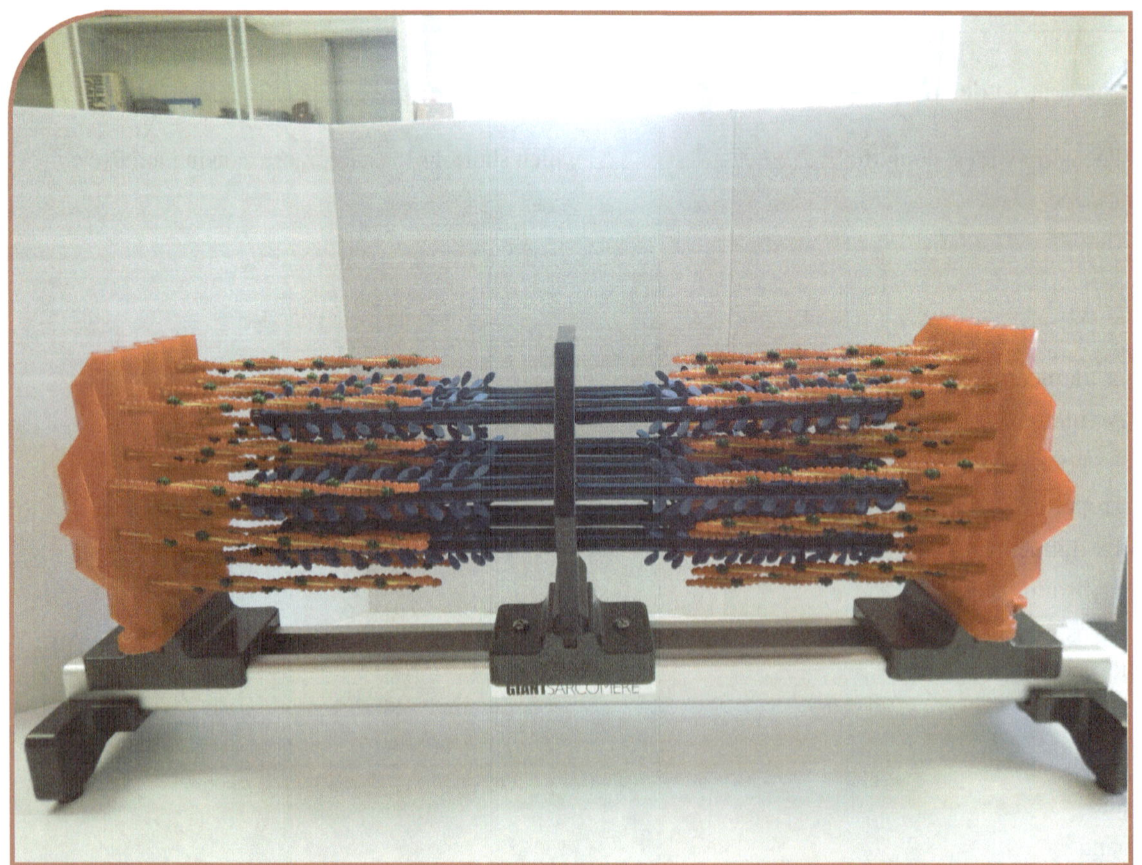

© Jodie Gerts, 2015

Muscle	Origin	Insertion	Action
Iliotibial band (not a muscle)	Mid, iliac crest	Lateral condyle of tibia	NO ACTION
Iliopsoas (iliacus part)	Anterior and posterior iliac fossa	Lesser trochanter	Flexion of the hip
Gluteus medius	Anterior (Not ASIS) and posterior iliac crest	Greater trochanter	Abduction of the hip
Gluteus maximus	Posterior iliac crest	Greater trochanter	Extension of hip
Tensor fasciae latae	Sternum and clavicle	Greater trochanter	Flexion and abduction of hip
Adductors	Pubis	Mid, medial femur	Adduction of hip
Quadriceps femoris	AIIS	Tibial tuberosity	Extension of knee
Hamstrings	Ischial tuberosity	Medial condyle of tibia and head of fibula	Flexion of knee
Tibialis anterior	Proximal, antiortibial shaft (just distal to tibial tuberosity)	Base of 1st metatarsal (Tendon wraps around medial malleolus)	Flexion of wrist
Gastrocnemius	Medial and lateral epicondyle of femur	Calcaneous	Plantarflexion of ankle
Soleus	Proximal posterior tibial shaft	Calcaneous	Plantarflexion of ankle
Fibularis longus	Lateral epicondyle of humerus	Head of fibula (Tendon wraps around lateral malleolus)	Eversion of foot
Intrinsic muscles of the toes	Originate in foot	Insert in foot (Plantar surface of calcaneous)	Toe movement (Planter surface of 1st Proximal phalanx)

BUILDING THE LEG MUSCLES

Line the pelvis up correctly with the head of the femur. Place the bands on the skeleton using the chart from above. Recommended order of placement based on the depth of the muscles. The deepest muscles will be placed first followed by the most superficial muscles. Pay attention to orientation of the bones, in particular the anterior and posterior side.

Tibialis anterior (2 clips)
    Wrap the tendon around the medial malleolus
Soleus (2 clips)
Intrinsic foot muscle (2 clips)

Fibularis longus (2 clips)
    Wrap the tendon around the lateral malleolus
Gastrocnemius (3 clips)
Iliopsoas (3 clips)
Gluteus medius (3 clips)

Adductors (2 clips)
Gluteus maximus (2 clips)
Tensor fasciae latae (2 clips)
Quadriceps (2 clips)
Hamstrings (3 clips)
Iliotibial band (2 clips)

RADIOGRAPHS

OTA Classification

The Orthopedic Trauma Association (OTA) uses a classification system to describe the fracture of long bones. There are minimum of five descriptors to document long bone fractures. Utilize these descriptors to classify fractures.

1. **Name the bone:** The OTA classification of a fracture starts by naming the bone involved, for example:

   A. Humerus fracture

   B. Radius fracture/Ulnar fracture

   C. Femoral fracture

2. **Location:** The part of the bone involved.

   A. Proximal

   B. Diaphyseal

   C. Distal

   D. Malleolar region of ankle is considered separately

3. **Type:** It is important to note the number of parts of the fracture and if it broke through the skin

   A. Simple or comminuted (multifragmentary)

   B. Closed or open

4. **Group:** The geometry of the fracture.

   A. Transverse

   B. Oblique/spiral

   C. (linear—we will not see examples of)

5. **Subgroup:** Other features of the fracture are described in terms of alignment

   A. Displacement—displaced or nondisplaced

   B. (Rotation, angulation, and shortening—we will not use this as an option)

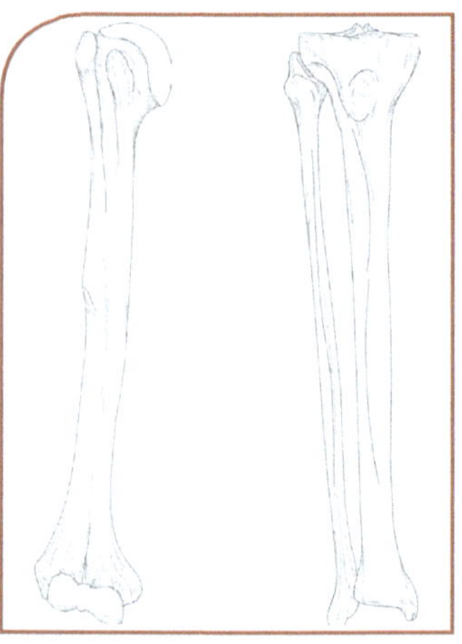

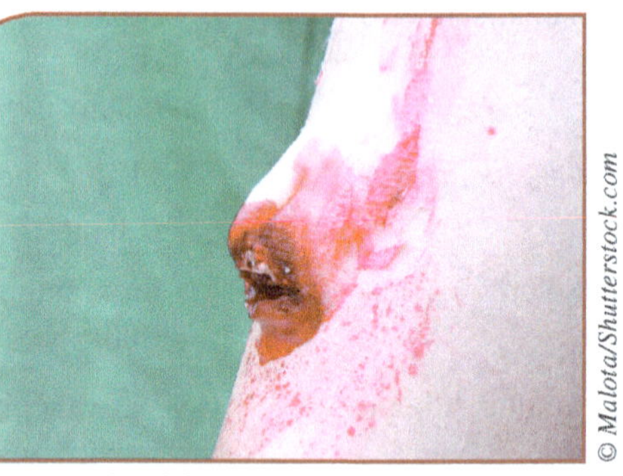

■ Example of Open Fracture

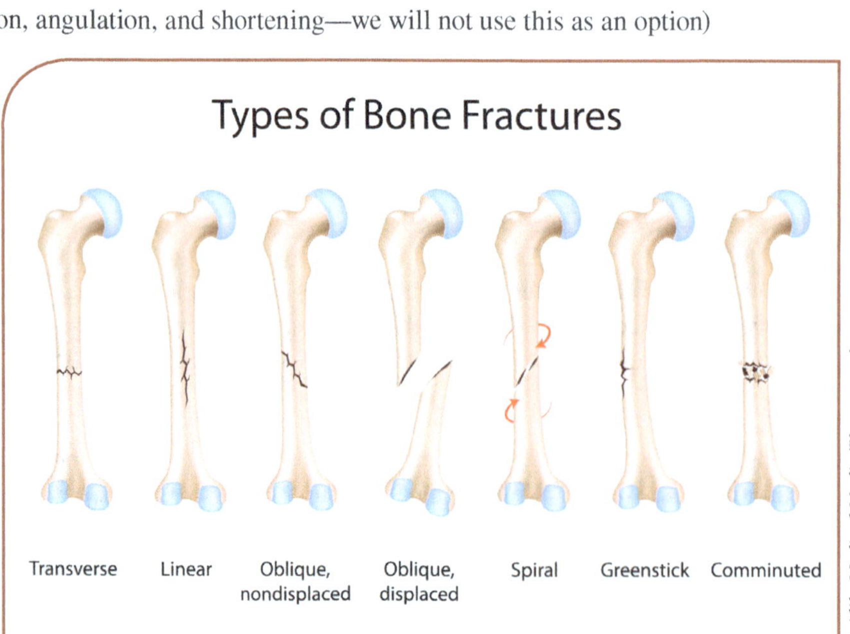

Expectations of the Student

1. **On an X-ray note:**

   A. Bone

   B. Location on bone (proximal, diaphyseal, distal, malleolar if it is a long bone)

   C. Type (transverse vs. spiral/oblique)

   D. Group (simple vs. comminuted and open vs. closed)

   E. Subgroup (displaced vs. nondisplaced)

2. **State type of treatment:** closed reduction non-invasive fixation, **open reduction internal fixation** (ORIF), closed reduction external fixation

   Closed reduction non-invasive fixation: Manipulation of the bone to bring the fracture ends in alignment without having to open the skin and using devices (i.e. cast) that do not break the skin to maintain the alignment.

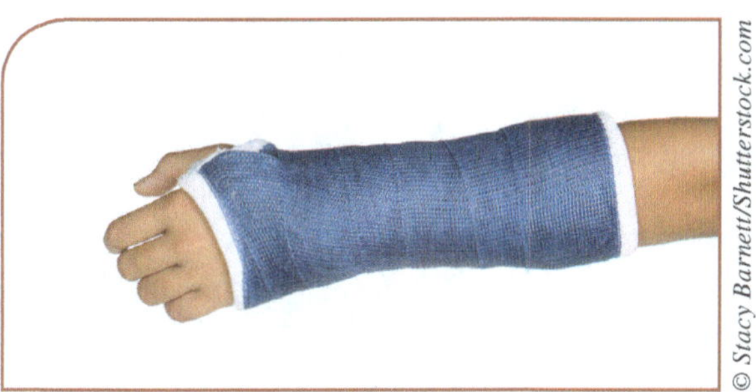

© Stacy Barnett/Shutterstock.com

   ORIF (open reduction internal fixation): Open reduction is when surgery is required to align the fracture. Internal fixation is using implants (i.e. screws) to maintain the alignment.

   Closed reduction external fixation: Using a device outside the skin that screws into the bone fragments to maintain bone alignment.

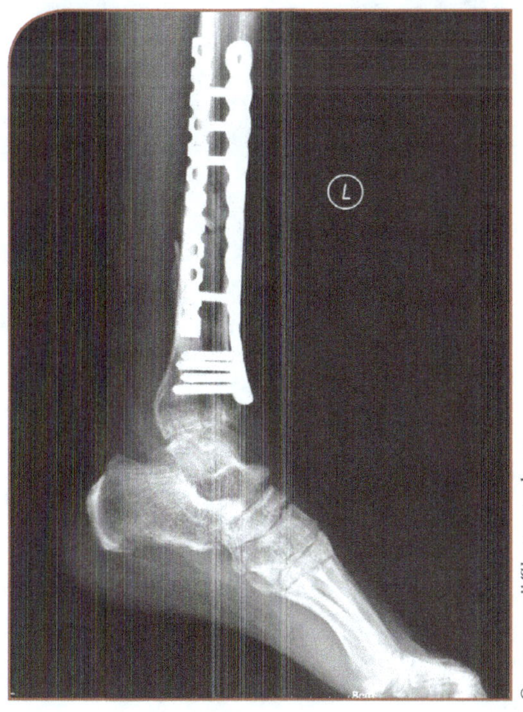

© sanyanwuji/Shutterstock.com

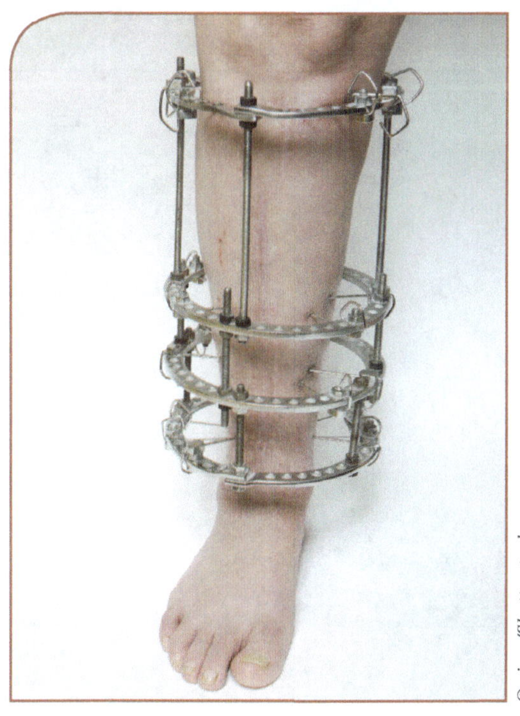

© sima/Shutterstock.com

LAB *8*

The Muscular System— Trunk and Upper Extremity

Objectives

The purpose of this lab is to have the student know the anatomy and function of the trunk, shoulder girdle and upper extremity muscles. The student will also utilize previously learned microanatomical structures to explain muscle physiology. The student will investigate muscle length-tension relationships using a hand dynamometer. The student will be able to perform strengthening and stretching exercises from knowing the origin, insertion and action of the muscles from Lab 7 and Lab 8.

Prior to lab: Complete the Pre-lab.

Complete APR Dissection, Histology, and Animations

During lab: Complete each objective prior to leaving lab.

☐ **Models**
   Skeletal Muscle Fiber
   Shoulder
   Elbow

☐ **Building the Trunk, Shoulder Girdle and Arm Muscles**

☐ **Length-Tension Relationship Experiment**

☐ **Perform Strengthening and Stretching Exercises**

LAB 8 PRELAB—The Muscular System—Trunk and Upper Extremity

Label these abdominal muscles on the picture. Color the origin in RED and color the insertion in BLUE:

Aponeurosis or anterior
rectus sheath

External oblique

Internal oblique

Linea alba

Rectus abdominis

Transverse abdominis

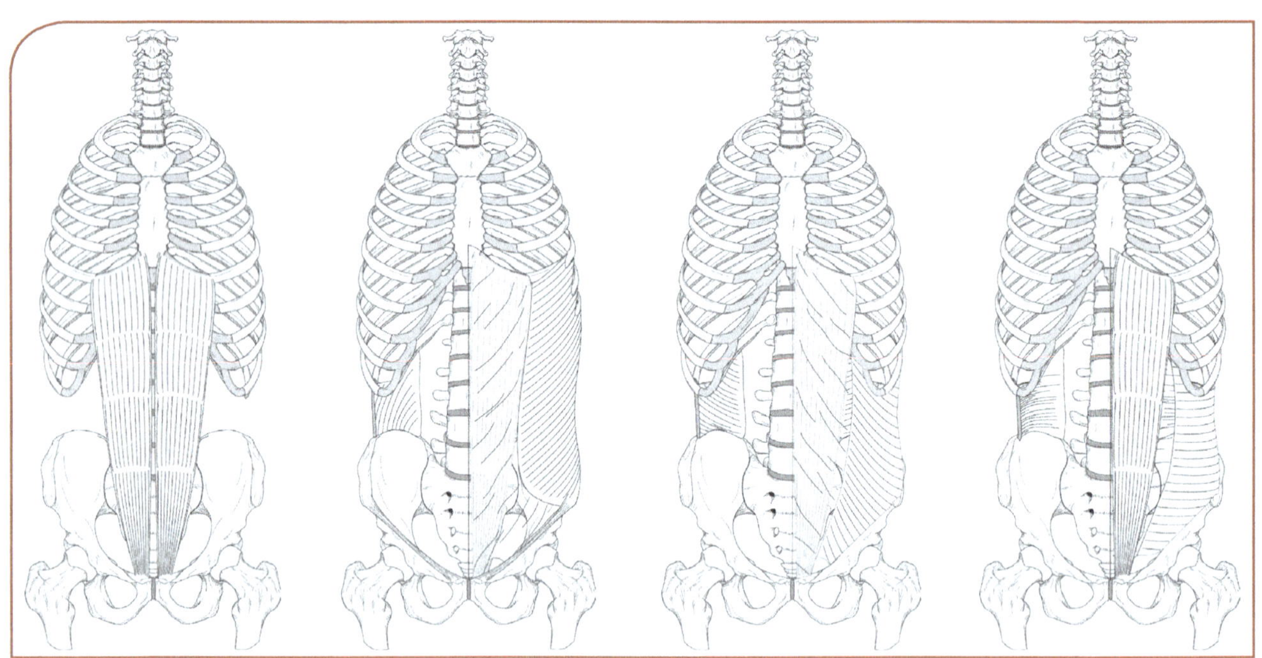

Illustrations by Jamey Garbett. © 2003 Mark Nielsen

Label these thoracic breathing muscles on the picture. Color the origin in RED and color the insertion in BLUE:

Diaphragm

Intercostals

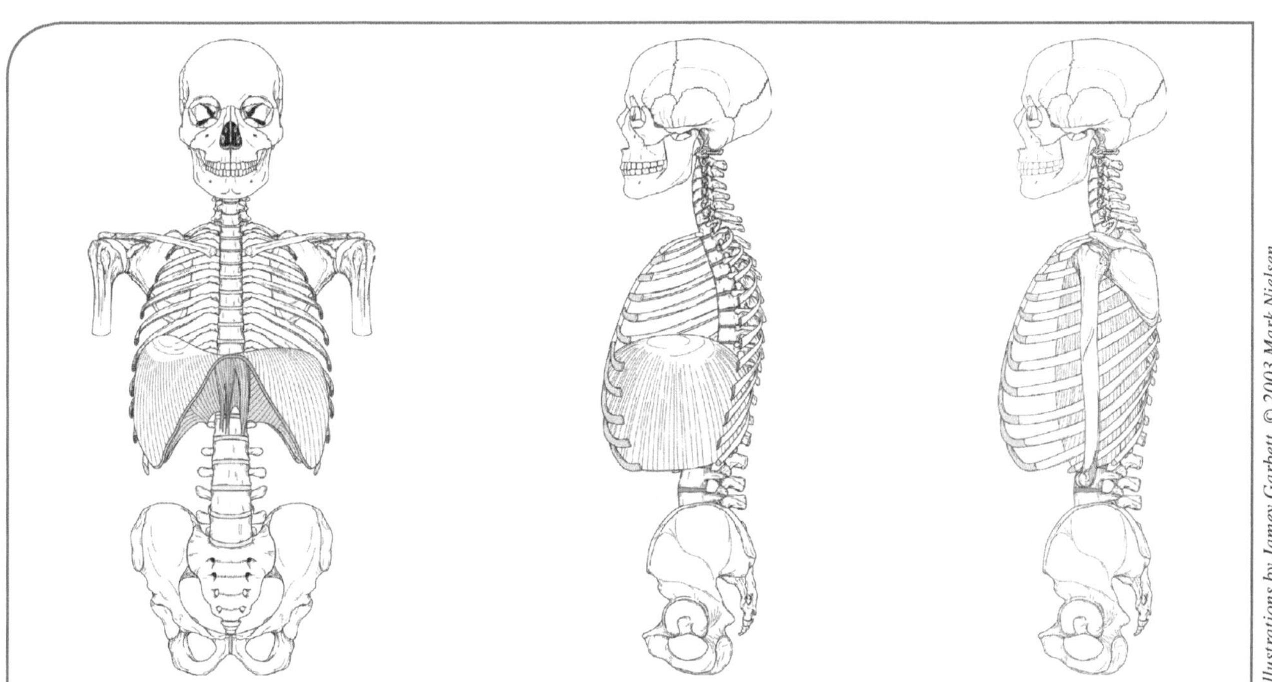

Label these rotator cuff muscles on the picture. Color the origin in RED and color the insertion in BLUE:

Infraspinatus

Supraspinatus

Subscapularis

Teres minor

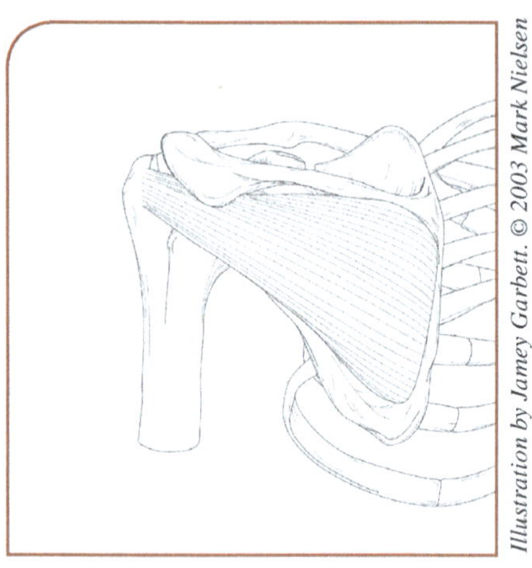

Illustration by Jamey Garbett. © 2003 Mark Nielsen

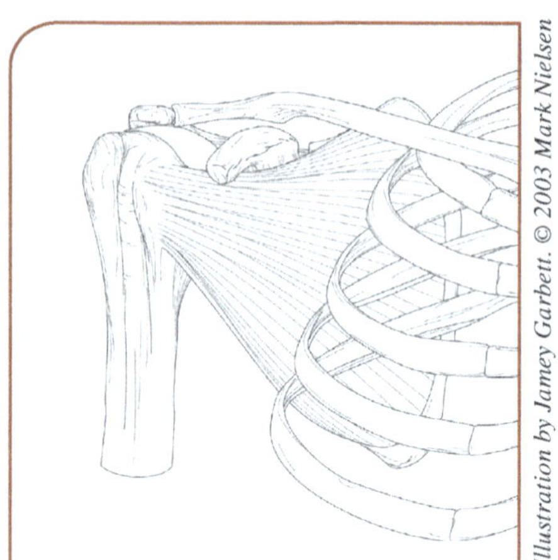

Illustration by Jamey Garbett. © 2003 Mark Nielsen

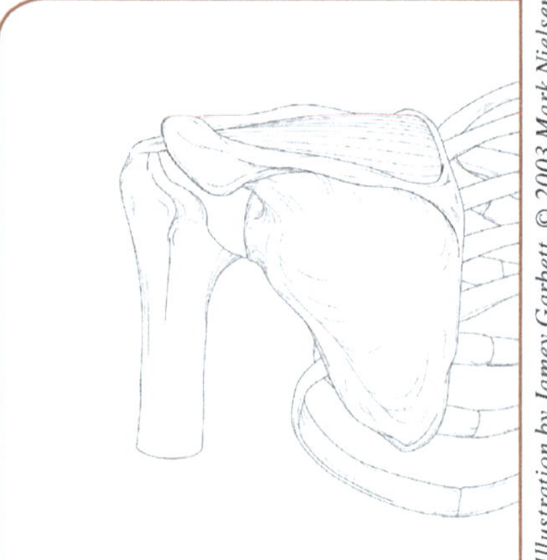

Illustration by Jamey Garbett. © 2003 Mark Nielsen

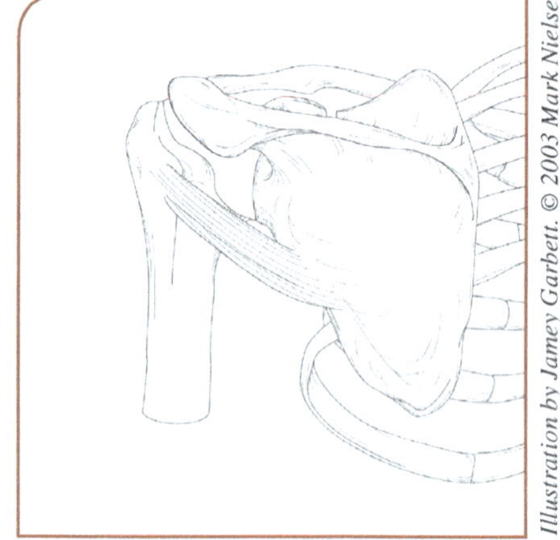

Illustration by Jamey Garbett. © 2003 Mark Nielsen

Label these back and scapular muscles on the picture. Color the origin in RED and color the insertion in BLUE:

Erector spinae

Rhomboid

Trapezius

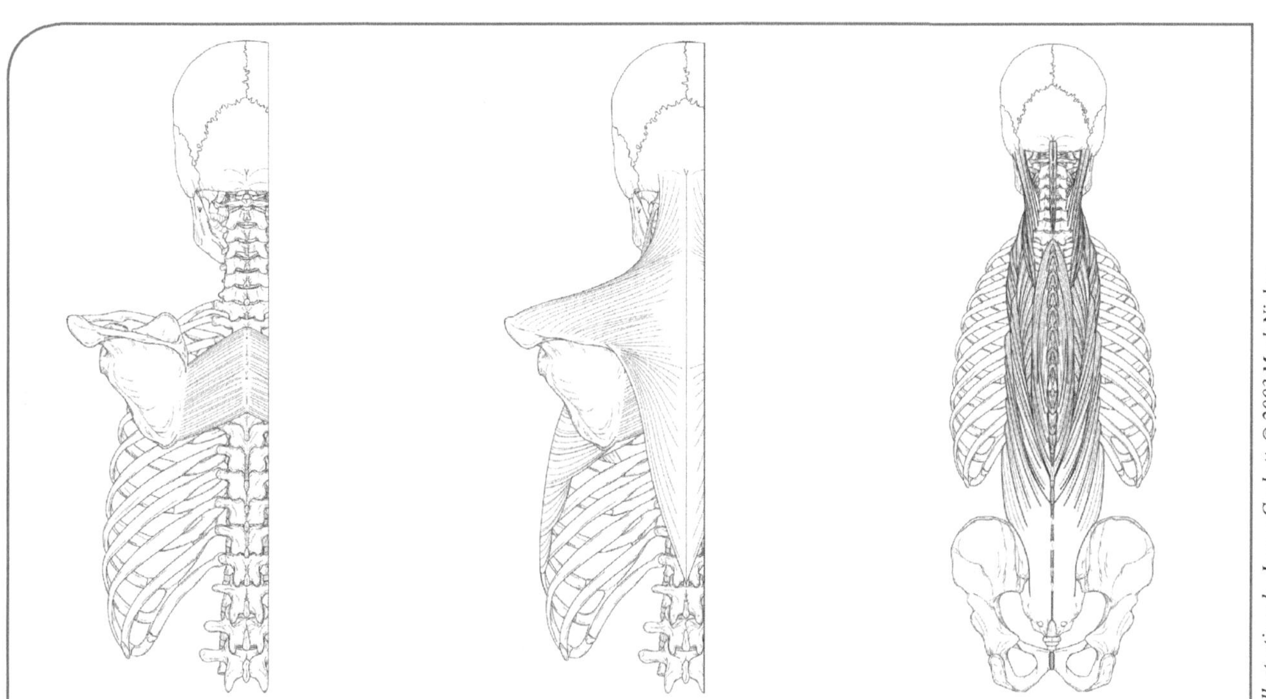

Label these shoulder and elbow muscles on the picture. Color the origin in RED and color the insertion in BLUE:

Biceps brachii	Latissimus dorsi
Brachialis	Pectoralis major
Deltoid	Triceps brachii

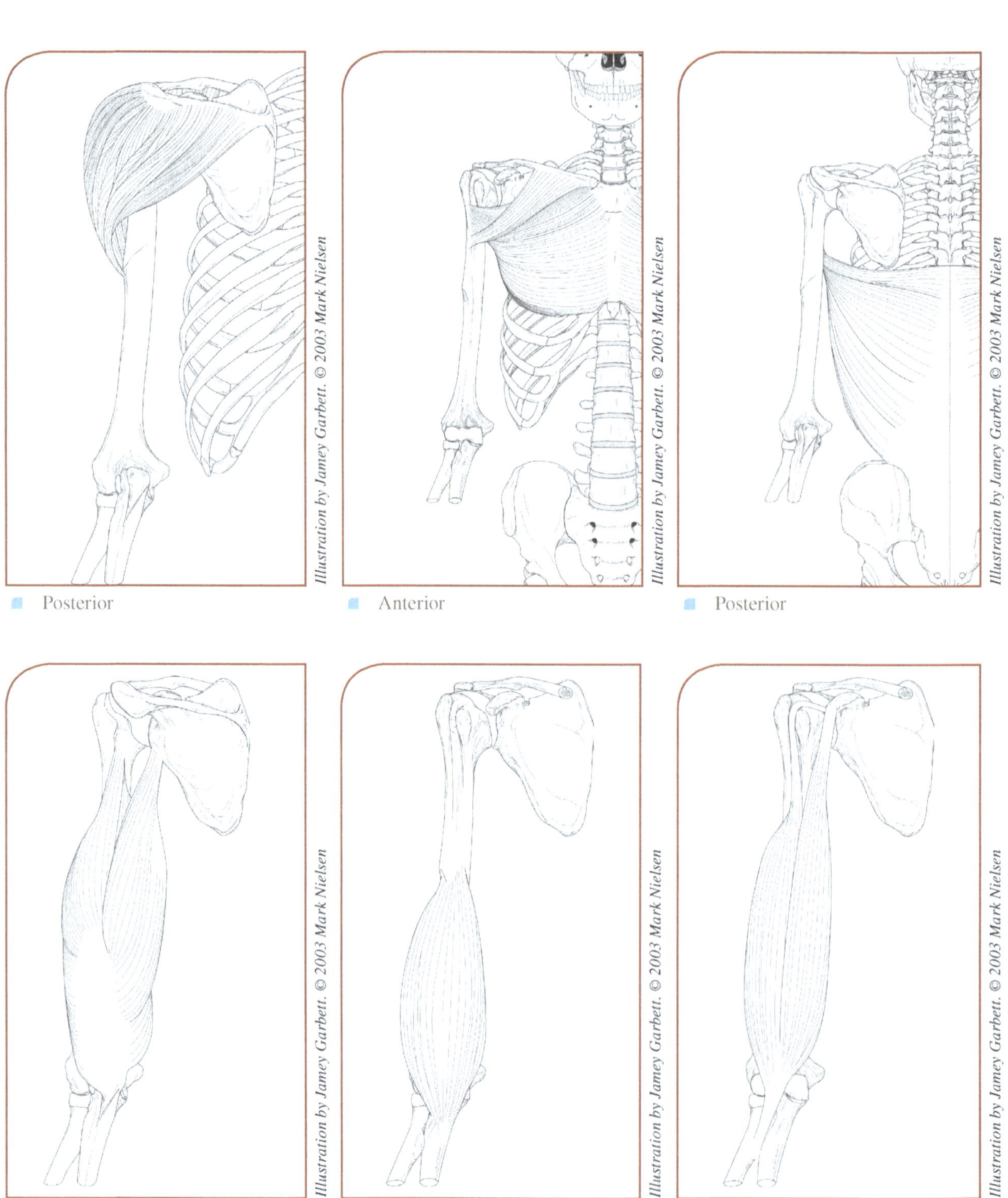

Posterior · Anterior · Posterior

Posterior · Anterior · Anterior

Illustration by Jamey Garbett. © 2003 Mark Nielsen

Label these forearm and hand muscles on the picture. Color the origin in RED and color the insertion in BLUE.

Extensors of the fingers   Flexors of the wrist

Extensors of the wrist   Intrinsics of the hand

Flexors of the fingers

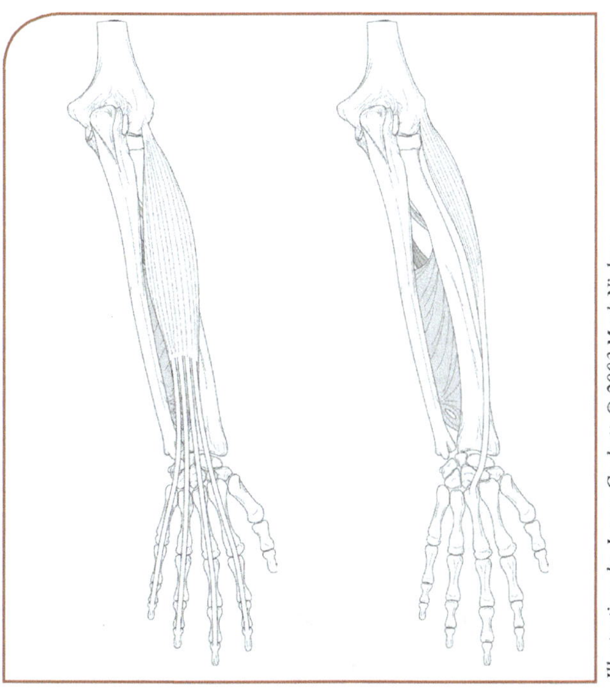

Posterior

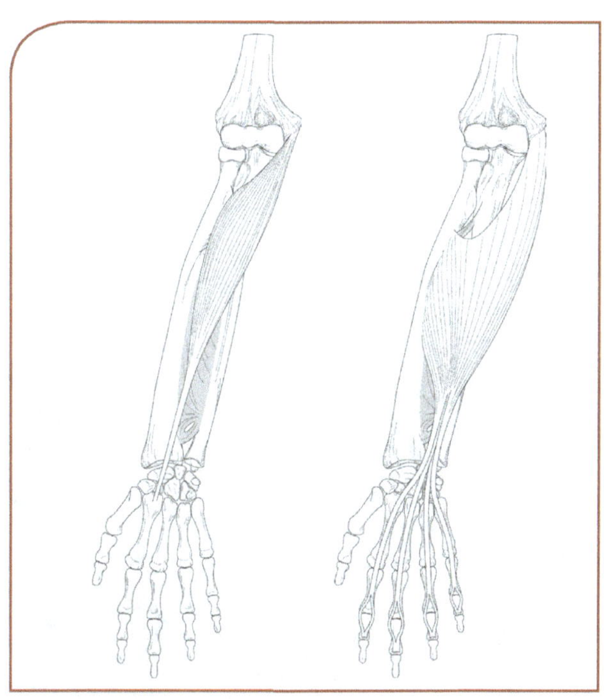

Anterior

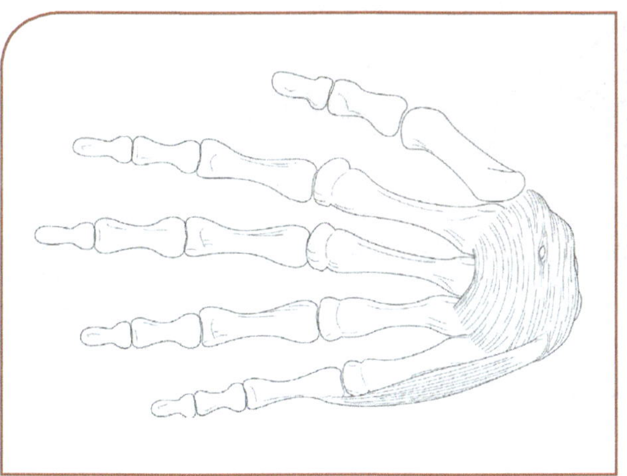

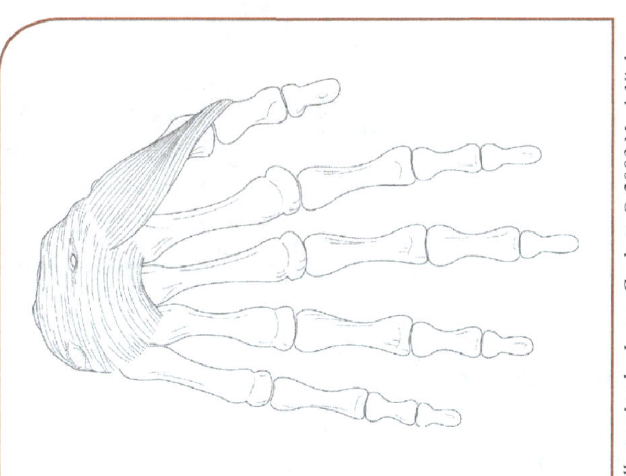

Identify the following structures in a neuromuscular junction:

Sarcolemma

Terminal knob or
synaptic knob

Acetylcholine vesicles

Acetylcholine

Motor end plate

Receptor for acetylcholine

Synaptic cleft

Sarcoplasm

Nerve axon

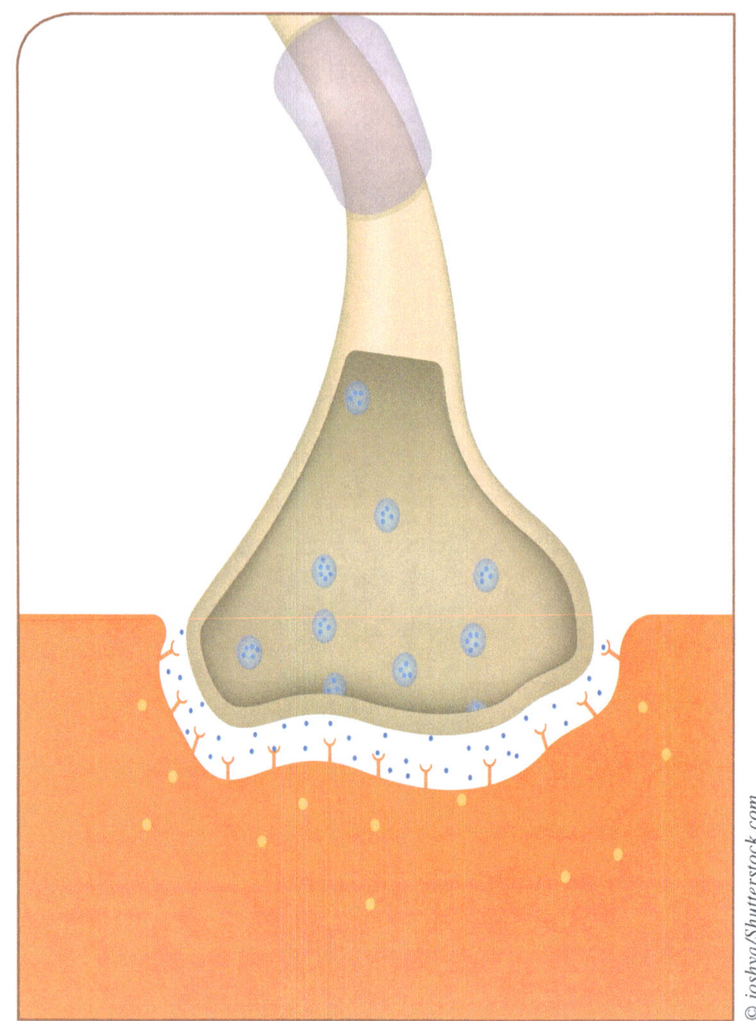

© joshya/Shutterstock.com

MODULE 6: THE MUSCULAR SYSTEM

Dissection

Head and Neck: Anterior

LAYER 3: Trapezius

Head and Neck: Lateral

LAYER 3: Trapezius

Head and Neck: Posterior

LAYER 2: Trapezius

Thorax: Anterior

LAYER 2: Anterior rectus sheath (can be called as aponeurosis), deltoid, latissimus dorsi, pectoralis major

What is an aponeurosis? ____

LAYER 3: External intercostal muscles (can be called intercostal muscle), rectus abdominis

LAYER 5: Diaphragm

Where does the diaphragm insert? ____

What is the primary muscle of inspiration? ____

Abdomen

LAYER 2: Anterior rectus sheath (can be called the aponeurosis), external abdominal oblique, inguinal ligament, linea alba, superficial inguinal ring

Where is a common site for male inguinal hernias? __________________________________

Where is the umbilicus located? ____

LAYER 3: Internal abdominal oblique, rectus abdominis

LAYER 4: Posterior rectus sheath (can be called an aponeurosis), transverse abdominis

Back

LAYER 2: Latissimus dorsi, trapezius

What bone does the latissimus dorsi insert on? _____________________________________

LAYER 3: Rhomboid

LAYER 5: Erector spinae muscles

Shoulder and Arm: Anterior

LAYER 2: Deltoid, pectoralis major

LAYER 3: Biceps brachii

What are the two primary actions of the biceps brachii muscle? _______________________________

LAYER 4: Brachialis, external abdominal oblique, subscapularis

How many muscles make up the rotator cuff? _______________________________

Shoulder and Arm: Posterior

LAYER 2: Deltoid, latissimus dorsi, trapezius

LAYER 3: Infraspinatus, rhomboid (major), teres minor, triceps brachii

LAYER 4: Supraspinatus

Forearm and Hand: Anterior

Many muscles are responsible for flexing and extending the wrist and fingers. In APR, we will use one muscle as an example of the many muscles that perform these actions.

Flexors of the wrist and flexors of the fingers muscle bellies are located on the medial surface of the forearm originating on the medial epicondyle of the humerus.

The extensors of the wrist and extensors of the fingers muscle bellies are located on the lateral surface of the forearm inserting on the lateral epicondyle of the humerus.

Muscles whose bellies are found within the hand and originate and insert on hand bones are intrinsic muscles of the hand.

LAYER 2: Flexor carpi radialis (can be called the flexors of wrist)

LAYER 3: Flexor digitorum superficialis (can be called the flexors of the fingers)

Forearm and Hand: Posterior

LAYER 3: Extensor carpi radialis brevis (can be called the extensors of wrist), extensor digitorum (can be called the extensors of fingers)

Wrist and Hand: Anterior

LAYER 3: Abductor digiti minimi (can be called intrinsic muscles of hand), abductor pollicis brevis (can be called intrinsic muscles of hand)

The muscles on the radial (thumb) side of the hand are called the thenar group.

The muscles on the ulnar (pinky) side of the hand are called the hypothenar group.

Histology: Neuromuscular Junction: LM High Magnification

Axon of motor endplate, motor end plate, skeletal muscle fiber

Is the axon to a skeletal muscle wrapped in myelin?   *YES   NO*   (circle)

Animation: Neuromuscular Junction

What type of ion channels open when an action potential reaches the presynaptic terminal?

____

When calcium moves into the presynaptic terminal, does it move up or down its concentration gradient?

____

What does calcium bind to one it enters the presynaptic terminal? _____________________________

What neurotransmitter is released from the presynaptic terminal? _____________________________

What type of ion channels does the neurotransmitter open? _____________________________

Movement of sodium into the muscle fiber causes   *repolarization   depolarization*   . (circle)

Does the neurotransmitter ever enter into the muscle fiber?   *YES   NO*   (circle)

What enzyme breaks down acetylcholine? _____________________________

Animation: Sliding Filament

In a muscle contraction, the _____________________ slides toward the center of the _____________________ filament.

When a muscle contracts, the _____________________ shorten.

The _____________________ disappears when the muscle is fully contracted.

Animation: Excitation-Contraction Coupling

Once an action potential is fired at the motor end plate of the neuromuscular junction, what

muscle structure is it propagated along? ___________________ And then is propagated down

the ________________.

The voltage change along the T tubules causes voltage-gated calcium ion channels to open on the

____

Calcium   *enters*   *leaves*   the sarcoplasmic reticulum.   (circle)

What proteins are attached to the tropomyosin? _______________________________________

What protein does calcium bind to? ____

And what effect does this cause? ____

What is uncovered when tropomyosin moves on the actin myofilament? ___________________

What is a cross-bridge? ____

Animation: Cross-Bridge Cycle

When the myosin heads bind to the myosin binding sites on actin, what is released from the myosin

heads? ____

The energy stored in the ________________ slides the actin past the myosin myofilament.

What breaks the bond between the myosin head and myosin binding site? _________________

ATP is broken down into ________________ and ________________ for energy.

Which ion must be present for the contraction to continue? ____________________________

SKELETAL MUSCLE FIBER MODEL

1. Thick filament (myosin)
2. Thin filament (actin)
4. Z disc
8. T tubule (transverse tubule)
9. Sarcoplasmic reticulum
10. Terminal cisternae of sarcoplasmic reticulum
11. Mitochondria
12. A nucleus of the muscle fiber
13. Sarcolemma
14. Opening of the T tubule
15. Synapse
16. Acetylcholine vesicle
17. Terminal nerve ending
18. Motor neuron axon
19. Myelin from Schwann cell

Motor end plate is the white part of the NMJ

Synaptic cleft is the red/pink part of the NMJ

Sarcomere—from Z line to Z line

> Differentiate between:
> Muscle fiber
> Myofibril
> Myofilament

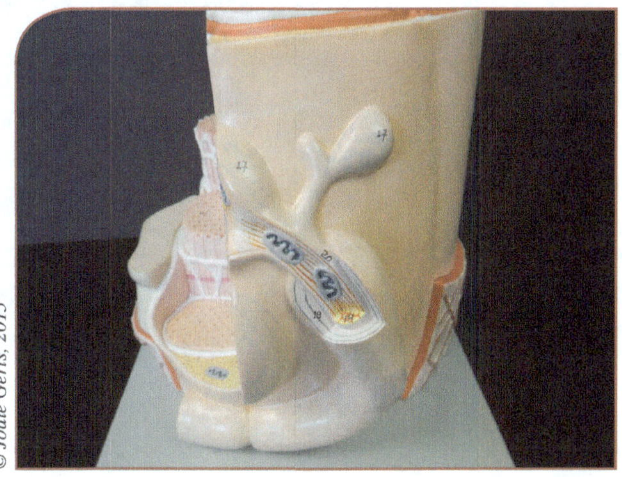

MODEL: ELBOW

Biceps brachii

Brachialis

Extensors of the wrist

Flexors of the wrist

Triceps

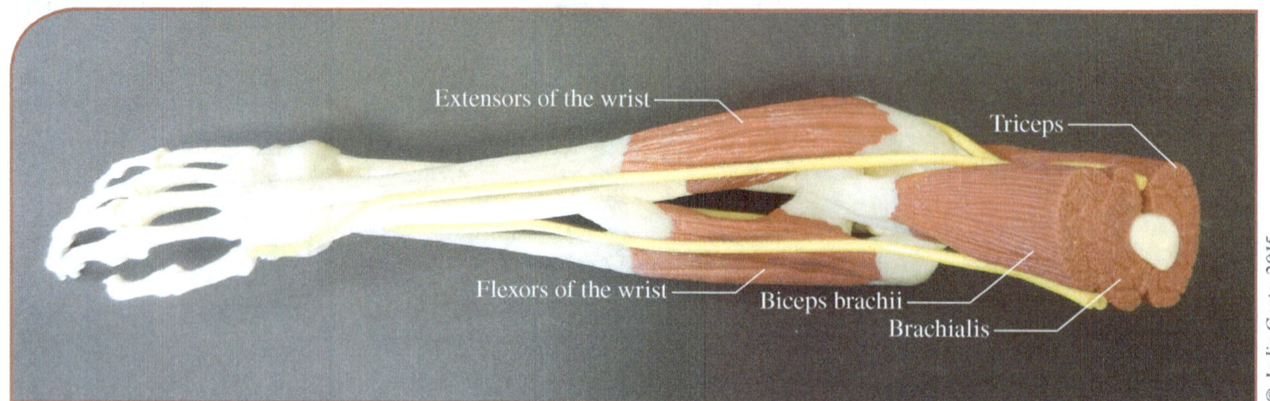

MODEL: SHOULDER

Supraspinatus	Subscapularis	Triceps
Infraspinatus	Latissimus dorsi	Biceps brachii
Teres minor	Deltoid	

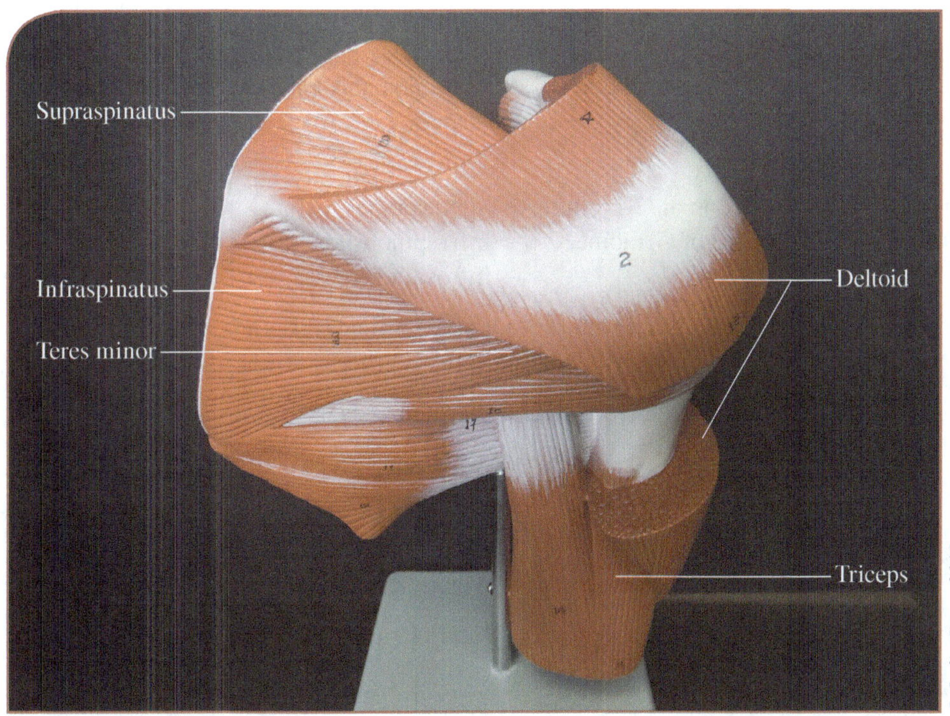

◢ Anterior View

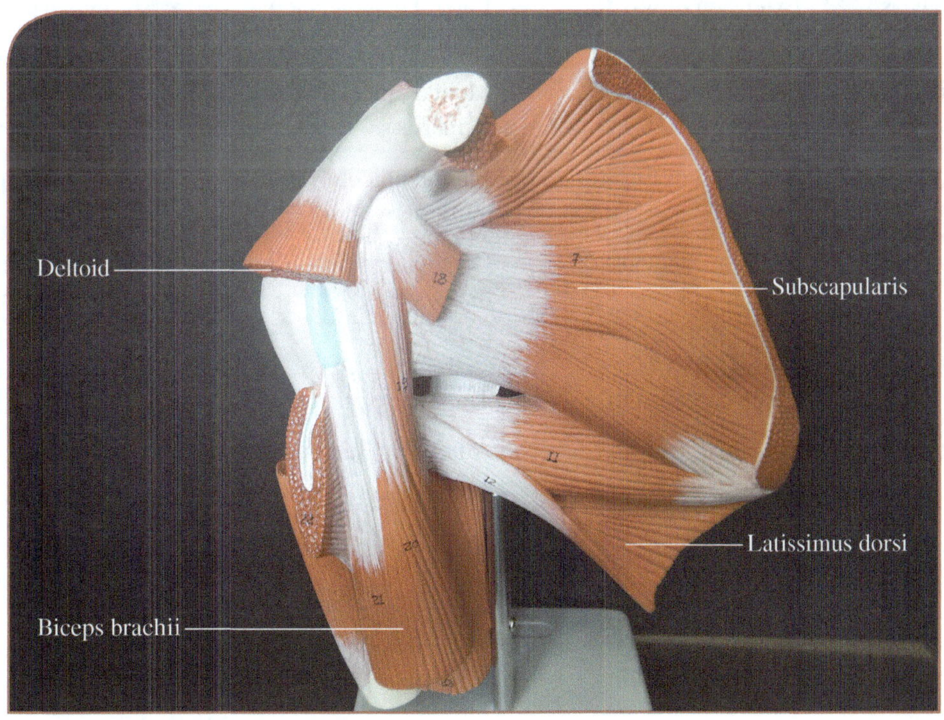

◢ Posterior View

Muscles of the Trunk, Shoulder Girdle and Upper Extremity

Muscle	Action	Origin	Insertion
Trapezius	Elevation, retraction of scapula	Spinous processes of vertebrae	Spine of scapula and acromion
Rhomboids	Retraction of scapula	Medial border of scapula	Spinous processes of upper thoracic vertebrae (T1 & T4)
Deltoid	Flexion and abduction of GH joint	Acromion and spine of scapula	Deltoid tuberosity of humerus

Rotator Cuff
    Supraspinatus
    Infraspinatus
    Teres minor
	Abduction of GH joint
	Rotation of GH joint
		Supraspinous fossa
		Infraspinous fossa
		Infraspinous fossa
			Greater tubercle
			Greater tubercle
			Greater tubercle

Latissimus dorsi	Adduction of GH joint	Posterior iliac crest and Spinous process of T11	Medial surgical neck of humerus
Pectoralis major	Adduction and IR of GH joint	Sternum and clavicle	Lateral surgical neck of humerus
Biceps brachii	Flexion and supination of forearm	Coracoid process	Radial tuberosity of radius
Brachialis	Flexion of the forearm	Anterior, mid humerus	Proximal, anterior ulna
Triceps brachii	Extension of elbow	Inferior rim of glenoid cavity	Olecranon
Flexors of the wrist	Flexion of wrist	Medial epicondyle of humerus	Anterior carpals
Extensors of the wrist	Extension of wrist	Lateral epicondyle of humerus	Posterior carpals
Flexors of the fingers	Flexion of fingers	Medial epicondyle of humerus	Anterior distal phalanx
Extensors of the fingers	Extension of fingers	Lateral epicondyle of humerus	Posterior distal phalanx
Intrinsics of the fingers	Finger movement	Originate in hand (Fifth metacarpal	Insert in hand (Fifth proximal phalanx)
Diaphragm	Inspiration	Ribs 10	Central tendon
Erector spinae	Extension of trunk	Transverse process of C3	Sacrum
Rectus abdominis	Flexion of trunk	Anterior costal cartilage	Pubic tubercles
External abdominal Oblique	Rotation of trunk	Lateral ribs 5 and 12	Pubis and anterior iliac crest
Internal abdominal oblique	Rotation of trunk	Anterior and posterior iliac crest	Anterior costal cartilage

BUILDING THE TRUNK, SHOULDER GIRDLE AND ARM MUSCLES

Using the red bands, place the muscles on the origin and insertion of the bones. You will place these muscles on the skeleton using the clips on the muscles. Use the previous chart to place the muscles on their respective approximate origins and insertions. Recommended order of placement based on the depth of the muscles. The deepest muscles will be placed first followed by the most superficial muscles.

1. Diaphragm (2 clips)
2. Brachialis (2 clips)
3. Erector spinae (2 clips)
4. Biceps brachii (2 clips)
5. Supraspinatus-thread it under the acromion (2 clips)
6. Triceps (2 clips)
7. Infraspinatus and teres minor (3 clips)
8. Rhomboids (4 clips)
9. Internal abdominal oblique (3 clips)
10. External abdominal oblique (4 clips)
11. Rectus abdominis (2 clips)
12. Latissimus dorsi (3 clips)
13. Trapezius (3 clips)
14. Pectoralis major (3 clips)
15. Deltoid (3 clips)
16. Flexors of the wrist (2 clips)
17. Extensors of the wrist (2 clips)
18. Flexors of the fingers (2 clips)
19. Extensors of the fingers (2 clips)
20. Intrinsics of the hand (2 clips)
21. Deltoid (3 clips)

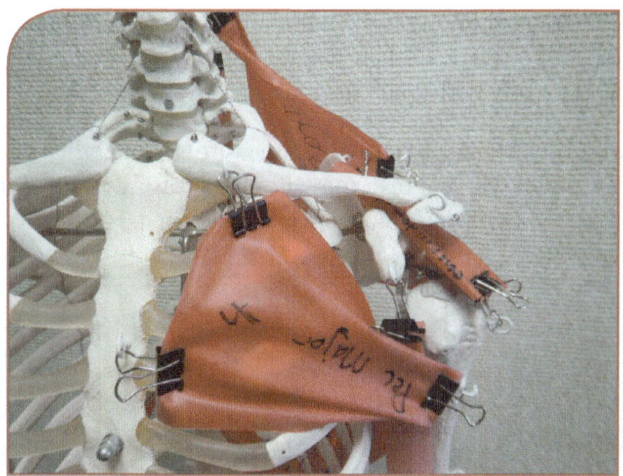

© Jodie Gerts, 2015

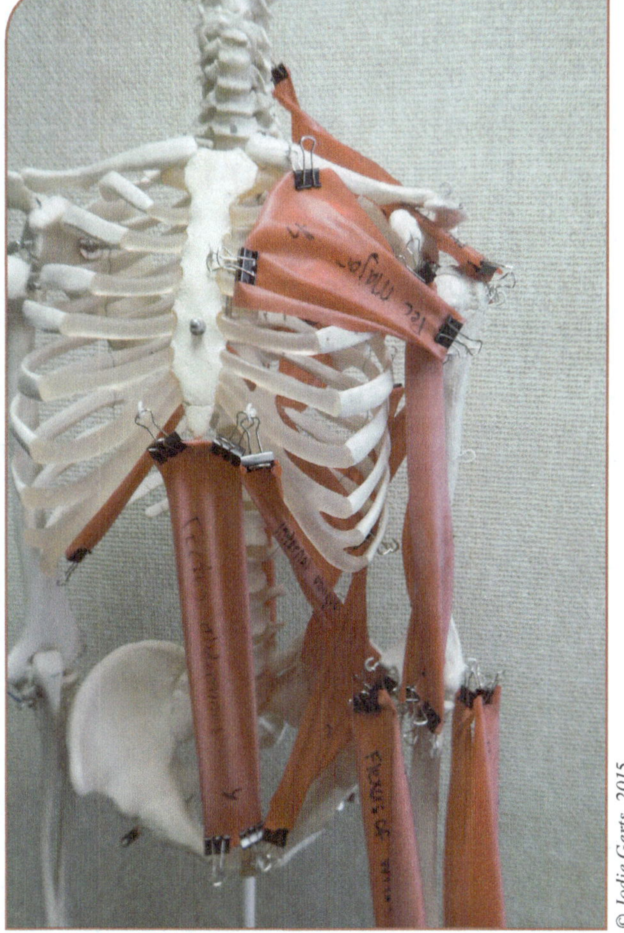

© Jodie Gerts, 2015

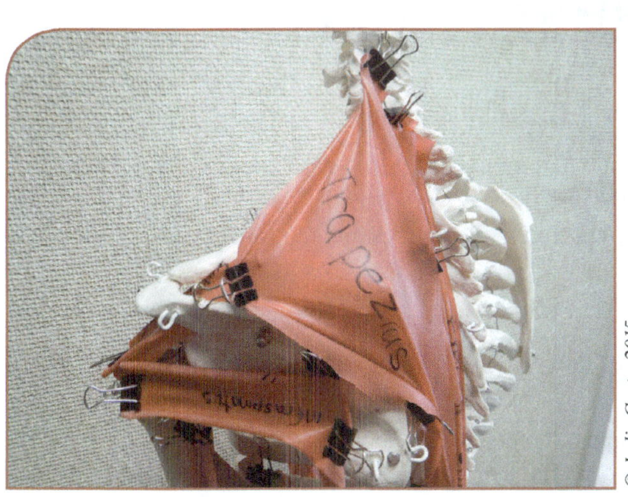

© Jodie Gerts, 2015

LAB ACTIVITY: Measuring Length-Tension Relationship

The amount of tension that can be generated during a contraction is influenced by how much the thick and thin filaments overlap when the muscle fiber is at rest—a relationship known as the length-tension relationship.

The overlap of the thick and thin filaments is determined by the length of the sarcomeres. When the muscle is stretched, the sarcomeres are longer and have less overlap. When the muscle is shortened, the sarcomeres are shorter and have more overlap.

The graph shows that the muscle fiber can generate the maximum amount of tension when its sarcomeres are at about 100–120% of their natural length. Conversely, the amount of tension generated during a contraction diminishes greatly both when the muscle is overly stretched or excessively shortened.

In this experiment you will be measuring the ability of your finger flexor muscles to generate tension at different resting muscle lengths. You will take measurements with an instrument called a hand dynamometer, which measures the force that you apply when the handle of the dynamometer is squeezed.

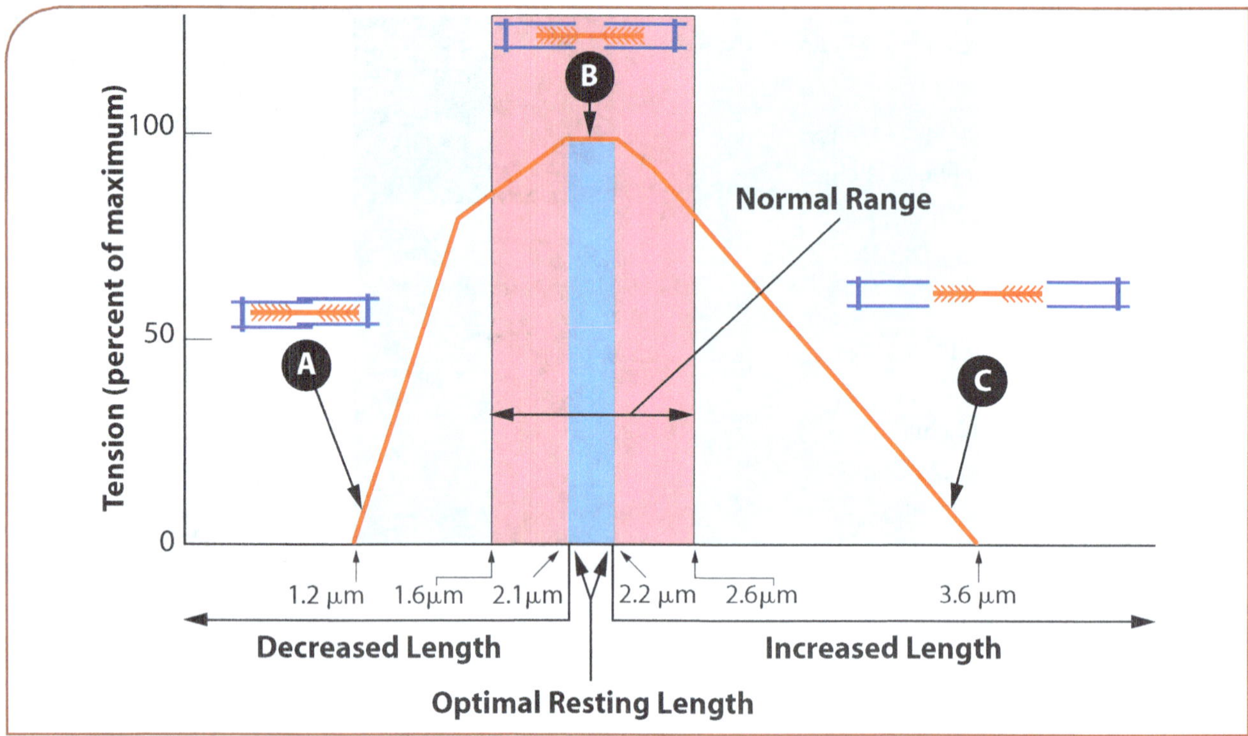

Testing with Wrist in Neutral

1. Set the handle to a comfortable grip and re-set max indicator to zero.
2. Position your dominant arm at your side with the elbow flexed to 90 degrees and the forearm in a neutral position.
3. Squeeze with maximum force and record reading.
4. Re-set to zero for next test.
5. Perform the test in 45 degrees flexion and record.
6. Perform the same test with the wrist in maximal flexion and record
7. Perform the test in 45 degrees extension and record.
8. Perform the test with the wrist in maximal extension and record.

Wrist Position	Amount of Force Generated (lb)
Full flexion (90)	
45 flexion	
Neutral	
45 extension	
Full extension (80)	

Interpret your results:

1. In which wrist position were the muscles' sarcomeres overly stretched?

2. In which wrist position were the muscles' sarcomeres excessively shortened?

3. Which wrist position yielded the most forceful contraction? Why?

4. Below is a table of the age-based norms for hand dynamometry for your dominant hand and gender.

Men				Women		
Mean	**Low**	**High**	**Age**	**Mean**	**Low**	**High**
121.0	91	167	**20–24**	70.4	46	95
120.8	78	158	**25–29**	74.5	48	97
121.8	70	170	**30–34**	78.7	46	137
119.7	76	176	**35–39**	74.1	50	99
116.8	84	165	**40–44**	70.4	38	103
109.9	65	155	**45–49**	62.2	39	100

Summary of Muscles from Labs 7 and 8

Muscles from APR	Muscles on Skeleton	Muscles to Know Action	Muscles to Know Origin and Insertion
Muscles of facial expression		Movement of face	
Extraocular muscles		Movement of eyes	
Temporalis		Mastication	
Masseter		Mastication	
Sternocleidomastoid		Flexion of head	
Trapezius	Trapezius	Elevation of scapula	
Rhomboids	Rhomboids		
Deltoid	Deltoid	Flexion and abduction of shoulder	o: Acromion and spine of scapula i: Deltoid tuberosity
Supraspinatus	Supraspinatus		
Infraspinatus	Infraspinatus	Rotation of shoulder	
Teres minor	Teres minor	Rotation of shoulder	
Subscapularis		Rotation of shoulder	
Latissimus dorsi	Latissimus dorsi	Extension of shoulder	
Pectoralis major	Pectoralis major	Flexion and adduction of shoulder	o: Coracoid process of scapula i: Radial tuberosity of radius
Biceps brachii	Biceps brachii	Flexion and supination of elbow	
Brachialis	Brachialis	Flexion of elbow	o: Glenoid cavity of scapula i: Olecranon

Summary of Muscles from Labs 7 and 8 *(continued)*

Muscles from APR	Muscles on Skeleton	Muscles to Know Action	Muscles to Know Origin and Insertion
Triceps brachii	Triceps brachii	Extension of elbow	
Flexors of the wrist	Flexors of the wrist	Flexion of wrist	
Extensors of the wrist	Extensors of the wrist	Extension of wrist	
Flexors of the fingers	Flexors of the fingers	Flexion of fingers	
Extensors of the fingers	Extensors of the fingers	Extension of fingers	
Intrinsics of the fingers	Intrinsics of the fingers		
Diaphragm	Diaphragm	Inspiration	o: Ribs 10 i: Central tendon
Intercostals			
Erecter spinae	Erecter spinae	Extension of spine	o: Costal cartilage i: Pubic tubercles
Rectus abdominis	Rectus abdominis	Flexion of spine	
External abdominal oblique	External abdominal oblique	Rotation of spine	
Internal abdominal oblique	Internal abdominal oblique	Rotation of spine	
Transverse abdominis			
Pelvic floor muscles			
Iliopsoas	Iliopsoas (iliacus part)	Flexion of hip	
Gluteus maximus	Gluteus maximus	Extension of hip	
Gluteus medius	Gluteus medius	Abduction of hip	
Piriformis			
Tensor fascia latae	Tensor fascia latae		
Hamstrings group	Hamstrings group	Flexion of knee	o: Ischial tuberosity i: Medial tibial condyle and fibular head
Quadriceps group	Quadriceps group	Extension of knee	o: AIIS i: Tibial tuberosity
Sartorius	Sartorius		
Adductor muscles group	Adductor muscle group	Adduction of hip	
Tibialis anterior	Tibialis anterior	Dorsiflexion of ankle	
Fibularislongus	Fibularislongus	Eversion of foot	
Gastrocnemius	Gastrocnemius	Plantarflexion of ankle	o: Medial and lateral epicondyles of femur i: Calcaneous

Summary of Muscles from Labs 7 and 8 *(continued)*			
Muscles from APR	**Muscles on Skeleton**	**Muscles to Know Action**	**Muscles to Know Origin and Insertion**
Soleus	Soleus	Plantarflexion of ankle	
Flexors of the toes		Flexion of toes	
Extensors of the toes		Extension of toes	
Intrinsic muscle of foot	Intrinsic muscle of foot		

Stretches (move the origin and insertion further from each other)

1. Quadriceps
2. Gastrocnemius and soleus
3. Erector spinae group
4. Hamstrings
5. Latissimus dorsi
6. Extensors of the toes
7. Adductor muscles
8. Iliopsoas
9. Triceps brachii
10. Flexors of the fingers

Strengthening (move the origin and insertion closer to each by contracting the muscle)

1. Quadriceps
2. Gastrocnemius and soleus
3. Rectus abdominis
4. Gluteus medius
5. Gluteus maximus
6. Hamstrings
7. Fibularis longus
8. Tibialis anterior
9. Flexors of the toes
10. Biceps brachii
11. Deltoid
12. Trapezius
13. Flexors of the fingers

L A B 9

The Nervous System—Brain

Objectives

The purpose of this lab is to have the student know the structures and functions of the brain. The student will learn how to perform a dissection using the sheep brain and be able to label the parts of the brain. The student will also be able to identify the functional parts of the brain including the meninges and ventricles. The student will have a preliminary introduction to cranial nerves.

Prior to lab: Complete the Pre-lab.

Complete APR Dissection and Animations

During lab: Complete each objective prior to leaving lab.

☐ **Models**

Meninges

Ventricles

Sagittal Section of the Head

☐ **Human Brain Specimen**

☐ **Identification of Parts of the Brain**

☐ **Brain Dissection**

☐ **Know Names and Numbers of Cranial Nerves**

LAB 9 PRELAB—The Nervous System—Brain

Identify the following structures of the ventricles:

1. Lateral ventricles
2. 3rd ventricle
3. 4th ventricle
4. Interventricular foramina
5. Cerebral aqueduct
6. Central canal
7. Choroid plexus

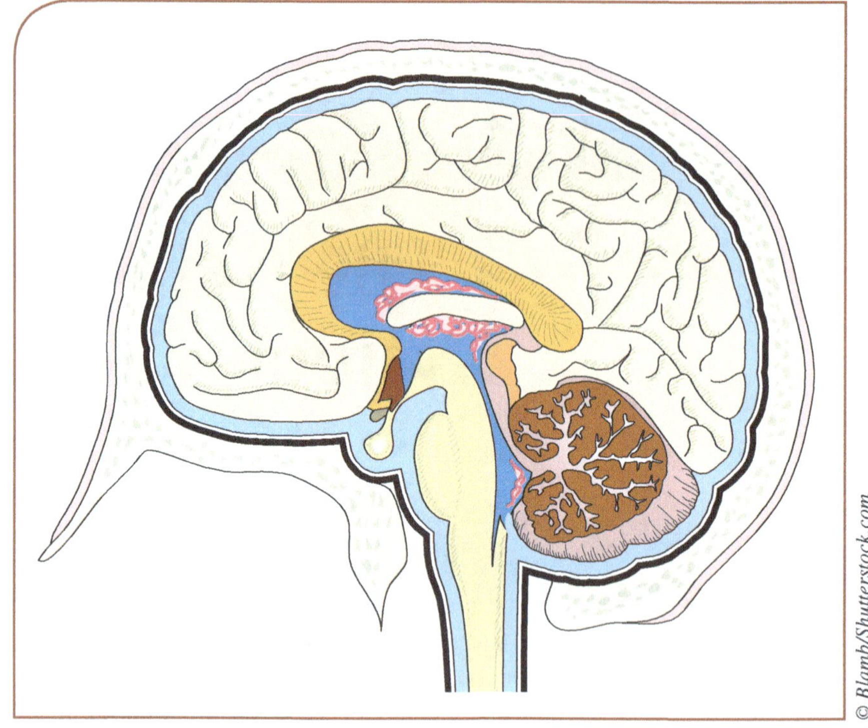

© Blamb/Shutterstock.com

Identify the following structures of the lateral brain:

1. Precentral gyrus (primary motor cortex)
2. Premotor cortex
3. Postcentral gyrus (primary somatosensory cortex)
4. Somatosensory association area
5. Primary visual cortex
6. Visual association area
7. Auditory cortex
8. Auditory association area
9. Broca's area (Motor speech cortex)
10. Wernicke's area
11. Prefrontal cortex
12. Taste area (Gustatory)
13. Central sulcus
14. Cerebellum

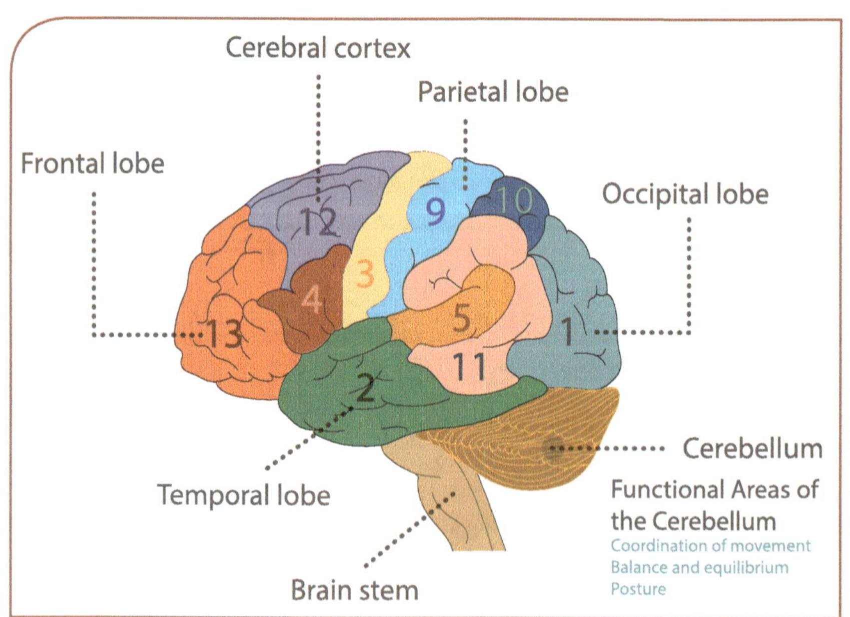

© Iaryna Turchyniak/Shutterstock.com

Identify the following structures of the sagittal brain:

1. Cerebellum
2. Arbor vitae
3. Midbrain
4. Pons
5. Medulla oblongata
6. Spinal cord
7. Hypothalamus
8. Thalamus
9. Third ventricle
10. Corpus callosum
11. Pineal gland
12. Pituitary gland
13. Optic chiasm
14. Frontal lobe
15. Parietal lobe
16. Occipital lobe
17. Cerebral aqueduct
18. 4th ventricle

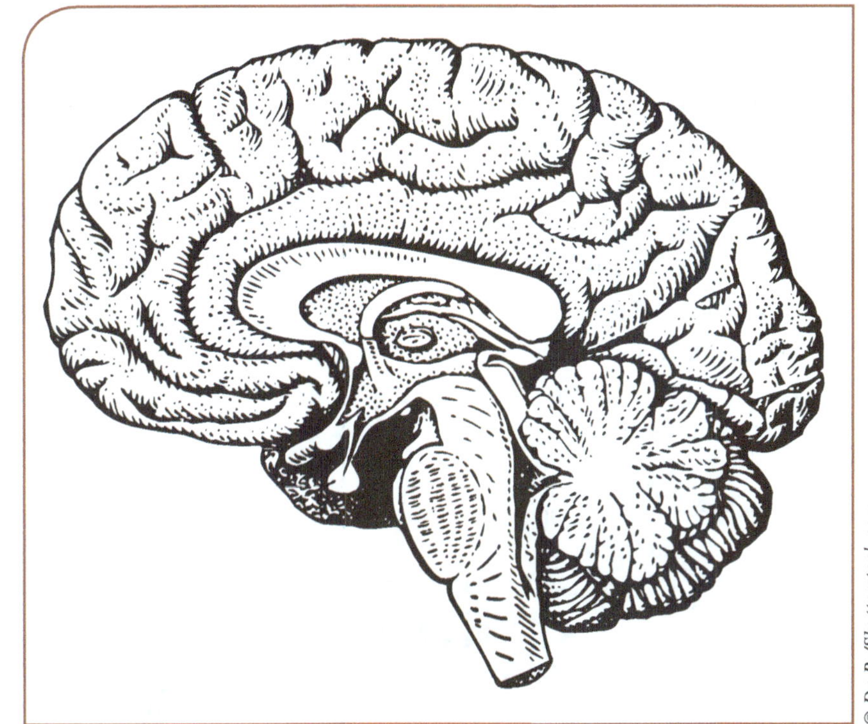

Identify the following structures of the inferior brain:

1. Olfactory tracts
2. Olfactory bulbs
3. Optic nerve
4. Optic chiasm
5. Optic tracts
6. Cranial nerve V
7. Cerebellum
8. Temporal lobe
9. Spinal cord
10. Pons
11. Medulla oblongata
12. Trigeminal nerve
13. Frontal lobe

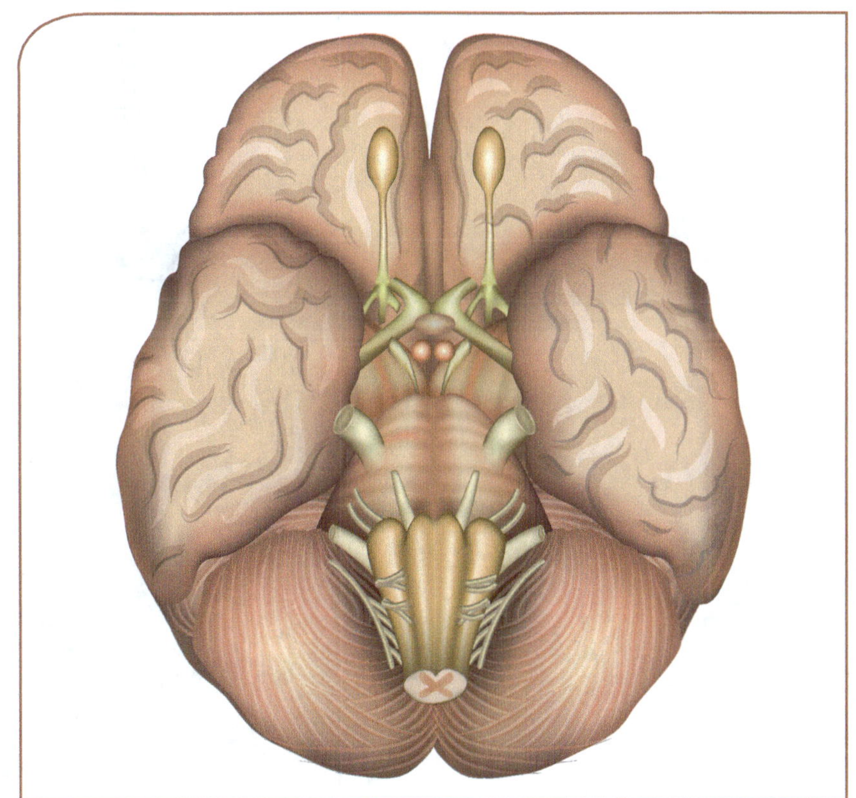

Identify the following structures of the brain meninges:

1. Skin
2. Skull
3. Dura mater
   - 3a. Periosteal layer
   - 3b. Meningeal layer
4. Dural venous sinus (with venous blood)
5. Arachnoid mater
6. Subarachnoid space
7. Arachnoid granulations or arachnoid villi
8. Pia mater
9. Cerebral cortex
10. White matter
11. Cerebrospinal fluid
12. Blood vessel
13. Gyrus
14. Sulcus
15. Subdural space

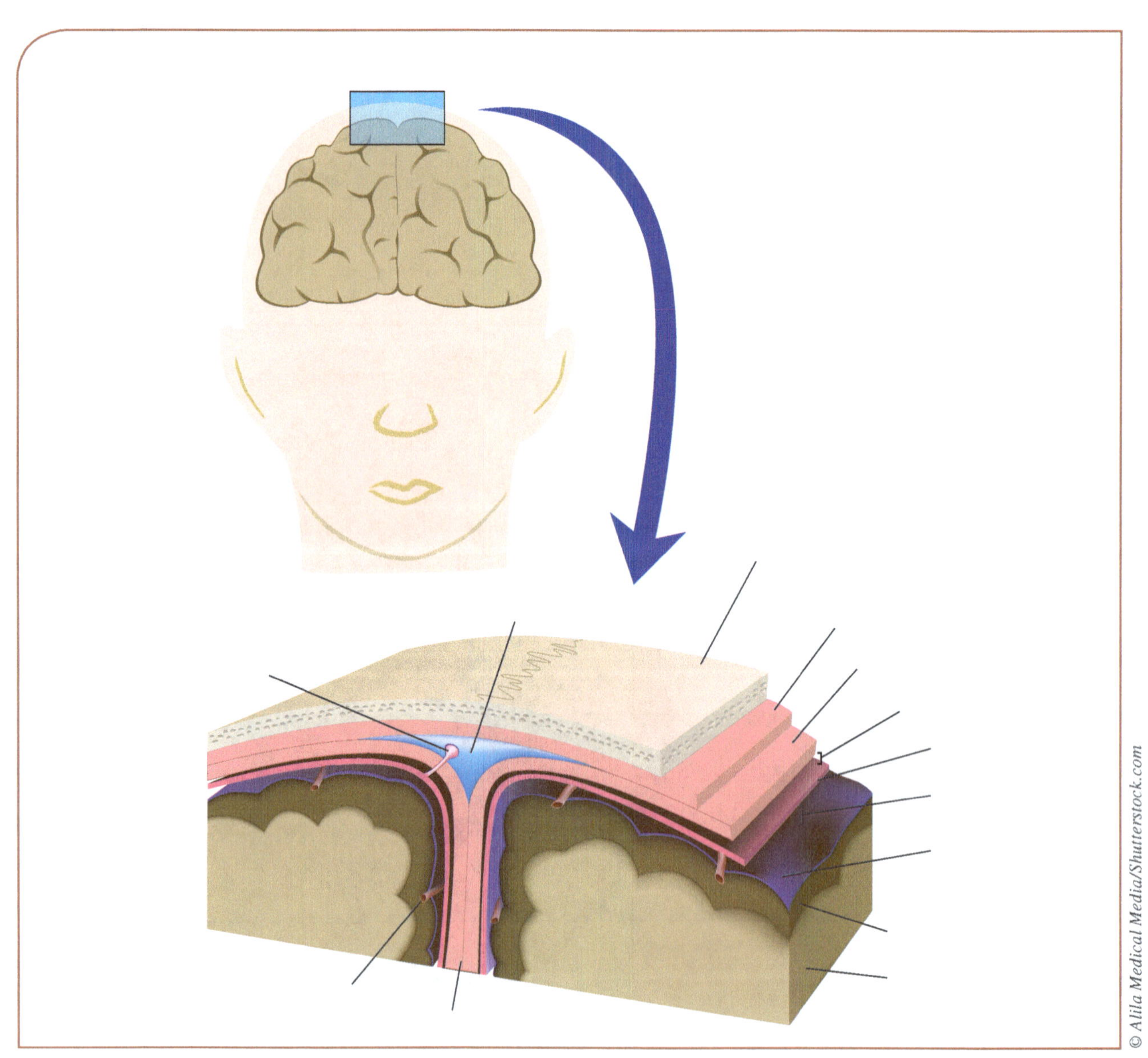

MODULE 7: NERVOUS SYSTEM

Dissection

Brain: Coronal

LAYER 1: Brainstem, cerebrum, frontal lobe, lateral sulcus, left cerebral hemisphere, longitudinal fissure, olfactory bulb, parietal lobe, temporal lobe

What fissure separates the cerebral hemispheres? ____

What lobe and what cerebral hemisphere is Wernicke's area located in? ____________________

What separates the temporal lobe from the frontal and parietal lobes? ____________________

LAYER 2: Gray matter, white matter

Which nervous tissue is unmyelinated? *Gray matter* *White matter* (circle)

LAYER 3: Corpus callosum, hippocampus, hypothalamus, lateral ventricle, thalamus, third ventricle

What is considered the master control center for the endocrine system? ____________________

What controls the autonomic nervous system? ____

What white matter structure connects the right and left hemisphere? ______________________

LAYER 4: Cerebral aqueduct, periaqueductal gray, red nucleus, substantia nigra

Loss of the neurons in the substancia nigra result in ________________ disease.

What part of the midbrain coordinates the movements of the eyes and head? ________________

LAYER 5: Cerebellum, corticospinal tract, fourth ventricle

What tract controls voluntary motor? ____

LAYER 6: Pyramidal tract

Brain: Inferior

LAYER 2: Cerebellum, cervical region of spinal cord, frontal lobe, infundibulum, medulla oblongata, occipital lobe, olfactory bulb, olfactory tract, optic chiasm, optic nerve, optic tract, pons, temporal lobe, trigeminal nerve

What part of the olfactory system synapses with the olfactory nerves? _____________________

Brain: Inferior Close-up

LAYER 2: Cervical spinal cord, infundibulum of pituitary gland, medulla, optic chiasm, optic nerve, optic tract, pons, trigeminal nerve

Brain: Lateral

LAYER 3: Dura mater

What is the function of the dura mater? ____

LAYER 4: Central sulcus, frontal lobe, lateral sulcus, occipital lobe, parietal lobe, precentral gyrus, postcentral gyrus, temporal lobe

Which lobe contains the postcentral gyrus? ____

Which lobe is the primary visual area? ____

The _________________ gyrus controls voluntary movment.

LAYER 6: Brainstem, cerebellum, cerebral aqueduct, cervical region of the spinal cord, corpus callosum, fourth ventricle, hypothalamus, left cerebral hemisphere, medulla oblongata, midbrain, pineal gland, pituitary gland, pons, thalamus, third ventricle

Brain: Superior

LAYER 3: Arachnoid granulations, dura mater

LAYER 4: Central sulcus, frontal lobe, parietal lobe, postcentral gyrus, precentral gyrus

LAYER 5: Basal nuclei, gray matter, gyrus, lateral ventricle, (not required to know horns), occipital lobe, sulcus, temporal lobe, thalamus, third ventricle, white matter

The basal nuclei are located   *deep   superficial*   in the brain.   (circle)

Animation: Divisions of the Brain

The cerebrum has _________________ large hemispheres.

Gyri are _________________ and sulci are _________________.

The four lobes of the cerebrum seen on the lateral surface of the brain are _________________,

_________________, _________________, _________________.

The cerebrum is responsible for higher levels of intelligence which includes information processing.

_________________, _________________ and _________________.

What part of the brain is responsible for coordinating motor activities? ______________________________

The thalamus is a part of the _____________________ and is a sensory relay station of the brain.

The hypothalamus is also part of the diencephalon and has multiple functions. Name five of those functions.

___________________________   ___________________________   ___________________________

___________________________   ___________________________

The brainstem is divided into three parts which are the _____________________, _____________________,

_____________________.

Which part of the brainstem is continuous with the spinal cord? ______________________________

Animation: Brain Ventricles Fly-through

How many ventricles are in the brain? ______________________________

Name the ventricles. _____________________, _____________________, _____________________

Where are the lateral ventricles located? ______________________________

What connects the lateral ventricle to the third ventricle? ______________________________

The narrow part of the third ventricle that leads to the fourth ventricle is called the _____________________.

The fourth ventricle of the brain is located in the ______________________________.

Cerebrospinal fluid (CSF) is produced by the ______________________________.

Animation: Meninges

The most superficial layer of the meninges is called the _____________________, the middle layer is the

_____________________ and the deepest layer is the _____________________.

Which sublayer of the dura mater is adhered to the skull? ______________________________.

The sinuses that hold venous blood are called ______________________________.

The subarachnoid space that contains the CSF is between the ______________________________.

Animation: CSF Flow

The cells in the choroid plexus that make CSF are called ______________________________.

The barrier that is selectively permeable is called the ______________________________.

What part of the spinal cord does the CSF flow through? _______________________________________

Where does CSF get recycled? ____

Animation: Action Potential Generation

Which channels are closed when the cell membrane is at resting membrane potential?

____

Which gates open first when depolarization occurs? _____________________________________

What happens to the voltage in depolarization?   *Becomes more positive   Becomes more negative*
(circle)

Depolarization occurs because more   *sodium   potassium*   diffuse into the cell.   (circle)

What changes as the action potential moves from depolarization to repolarization? _______________

Once the potassium gates close, sodium-potassium pumps restore the _______________ potential.

Animation: Action Potential Propagation

An action potential can move up and down the axon.   *True   False*   (circle)

The action potential becomes _______________ with respect to the outside of the neuron.

Why does the axon potential move in only one direction? _________________________________

Animation: Chemical Synapse

As the action potential arrives at the synaptic knob of the neuron, voltage-gated _______________
channel open.

Calcium then diffuses   *into   out of*   the cell.   (circle)

What is contained in the synaptic vesicles? ____

As the acetylcholine diffuses across the synaptic cleft, it binds to receptor sites on _______________.

When these gates open, _______________ moves into the cell.

If the voltage reaches threshold in the postsynaptic cell, an _______________ will be produced.

Meninges Model

Skin

Skull

Dura mater

Dural venous sinus

Arachnoid mater

Subarachnoid space

Arachnoid granulations

Pia mater

Cerebral cortex

White matter

Subdural space

Ventricles

Lateral ventricles

3rd ventricle

4th ventricle

Cerebral aqueduct

Interventricular foramen

Choroid plexus (pink part)

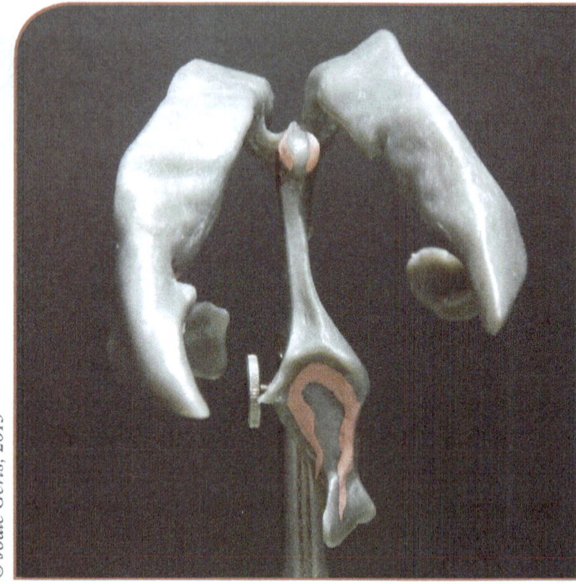

© Jodie Gerts, 2015

Sagittal Section of the Head

Cerebrum

Cerebellum

Midbrain

Pons

Medulla oblongata

Sinus

Arbor vitae

Corpus callosum

Thalamus

Cerebral aqueduct

4th ventricle

Pineal gland

Pituitary gland

Spinal cord

Hypothalamus

Human Brain

Frontal lobe

Parietal lobe

Temporal lobe

Occipital lobe

Pons

Medulla oblongata

Cerebellum

Optic chiasm

Olfactory bulb

Olfactory tract

Optic nerve

Optic tract

Trigeminal nerve

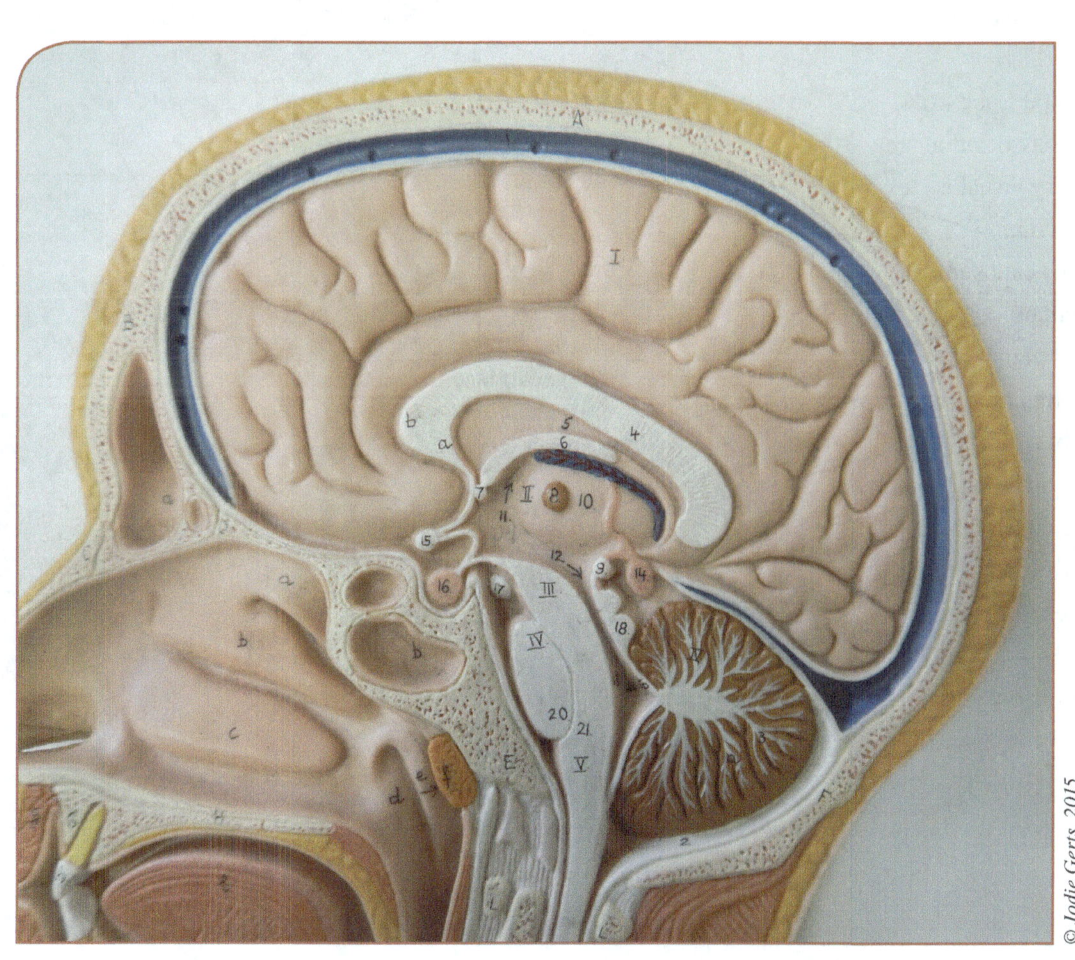

Brain Model

Identify and check off with instructor. Know general functions.

Cerebrum

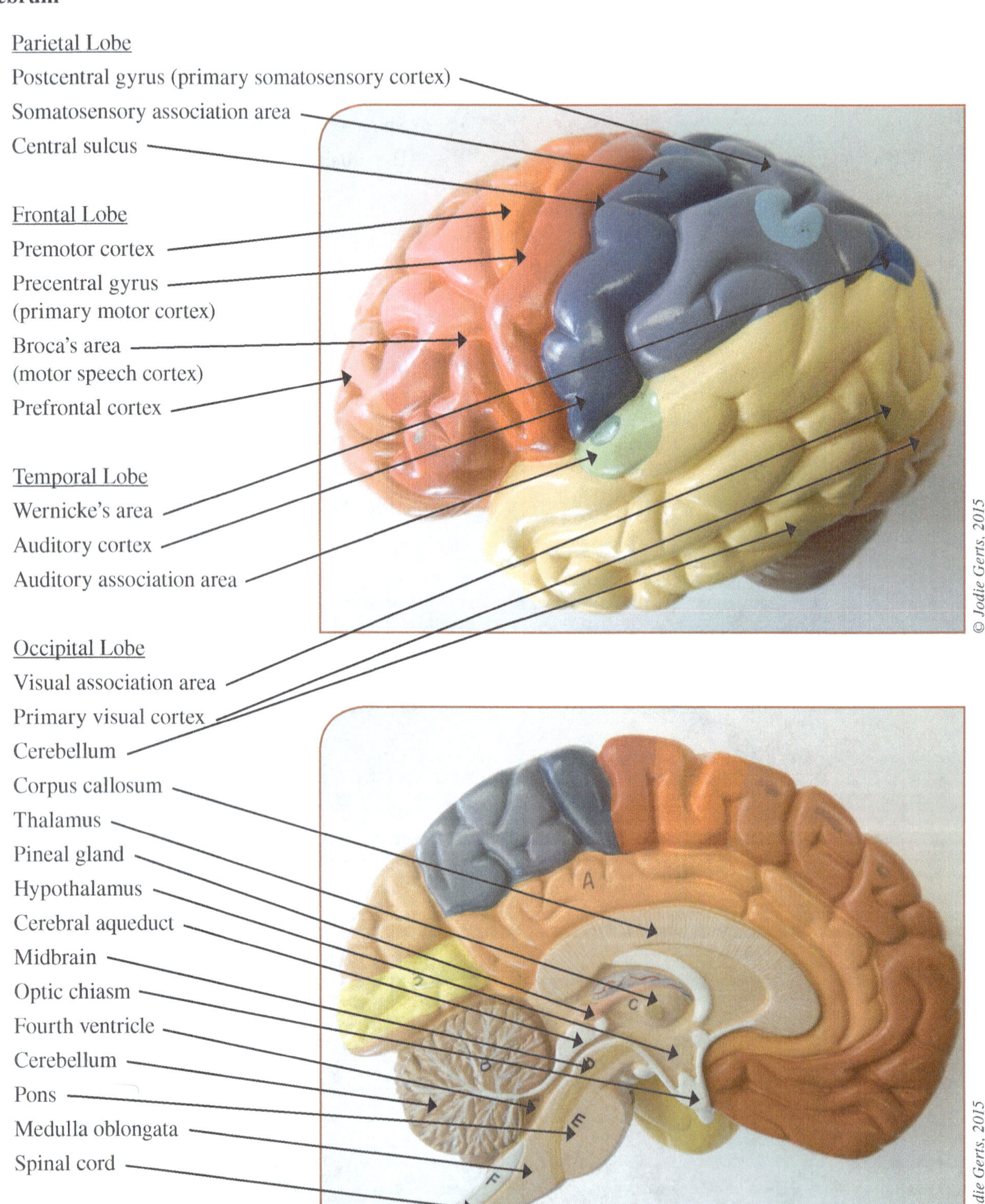

Parietal Lobe

Postcentral gyrus (primary somatosensory cortex)

Somatosensory association area

Central sulcus

Frontal Lobe

Premotor cortex

Precentral gyrus
(primary motor cortex)

Broca's area
(motor speech cortex)

Prefrontal cortex

Temporal Lobe

Wernicke's area

Auditory cortex

Auditory association area

Occipital Lobe

Visual association area

Primary visual cortex

Cerebellum

Corpus callosum

Thalamus

Pineal gland

Hypothalamus

Cerebral aqueduct

Midbrain

Optic chiasm

Fourth ventricle

Cerebellum

Pons

Medulla oblongata

Spinal cord

MAMMAL BRAIN DISSECTION GUIDE

Safety

Follow safe laboratory practices when performing any dissection. Perform dissections on a dissecting tray to contain specimens and fluids. Be careful when using sharp instruments such as scalpels, forceps, teasing needles, and scissors. Make sure you wear gloves. Goggles are provided for use as well.

Procedure

1. Rinse the brain in the sink and place brain on the dissecting tray.
2. Observe the dura mater. The two remaining meninges, pia and arachnoid, form a thin covering which adheres to the surface of the cerebellum. Use forceps and scissors to gently remove the dura mater.
3. Carefully dissect the optic nerve located within the orbital fat leading to the eye. Parts of the eye may still be intact.
4. Pin the structures identified previously listed under the whole brain. Use the pictures for assistance. Check off with your instructor and then remove the pins.
5. Place the brain on the dissecting tray with the longitudinal fissure side up. Using your fingers, gently widen the medial longitudinal fissure. Insert a scalpel into the fissure and cut through the corpus callosum connecting the two cerebral hemispheres. Continue to cut, perfectly dividing the cerebrum, cerebellum, brainstem, and cervical spinal cord into two longitudinal halves.
6. Identify and pin the structures listed under the sagittal cut section listed on the previous page. Check off with your instructor. Then remove the pins.
7. Place the brain on the dissecting tray and make a frontal cut using your scalpel through the midsection of the brain. This will expose the thalamus.
8. Identify and pin the structures listed under the frontal cut section on the previous page. Check off with your instructor. Remove the pins.
9. You are encouraged to make any further investigation at this point. Once you have observed all the structures of the brain, dispose of the specimen in the location indicated.
10. Remove your scalpel blade from the handle as shown in class.
11. Take your tray and instruments to the sink and wash with soap and water.
12. Dry all the equipment and place on the shelf so the instructor can prepare the specimens for the next class.

Pin the following structures of the brain. Start with the whole brain and check off with instructor and then pin the sagittal brain and check off with instructor.

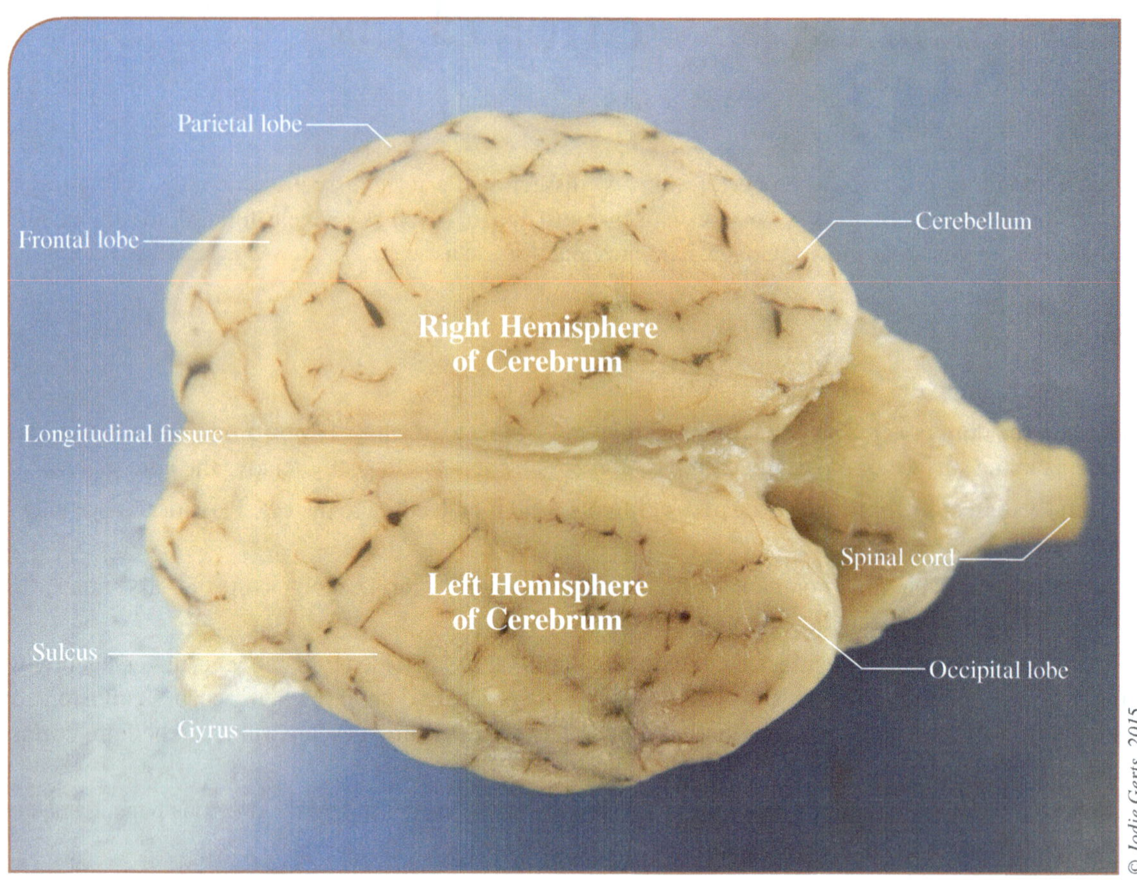

Superior View

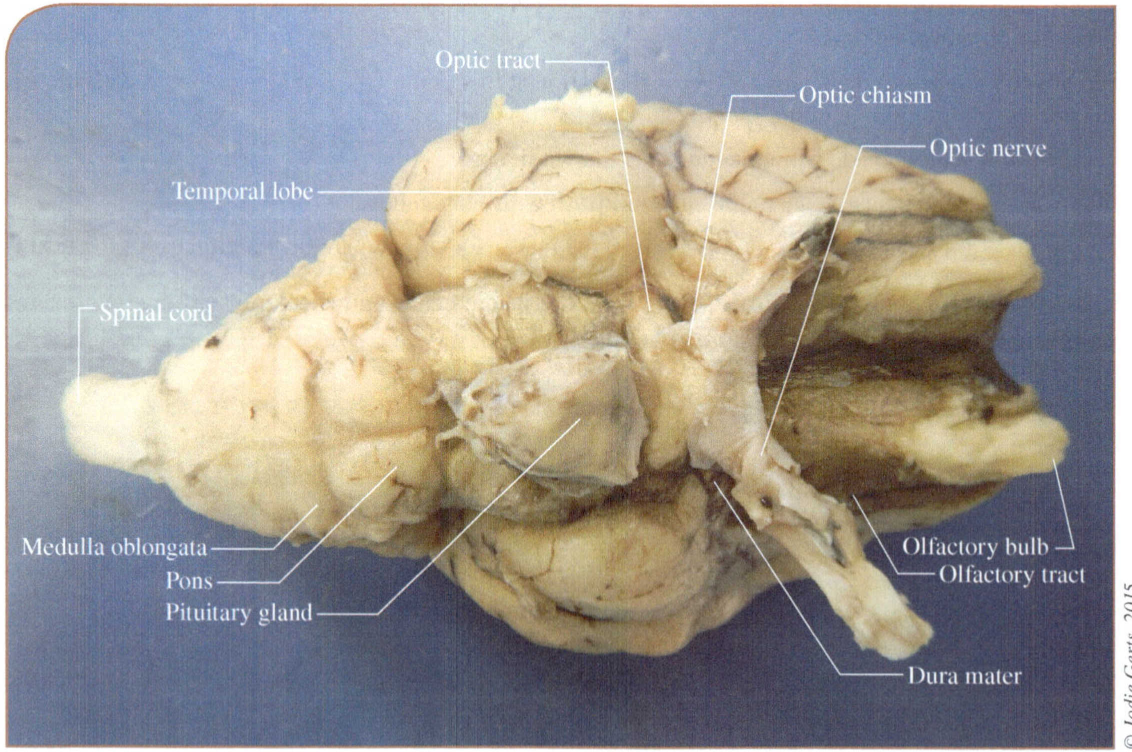

Inferior View

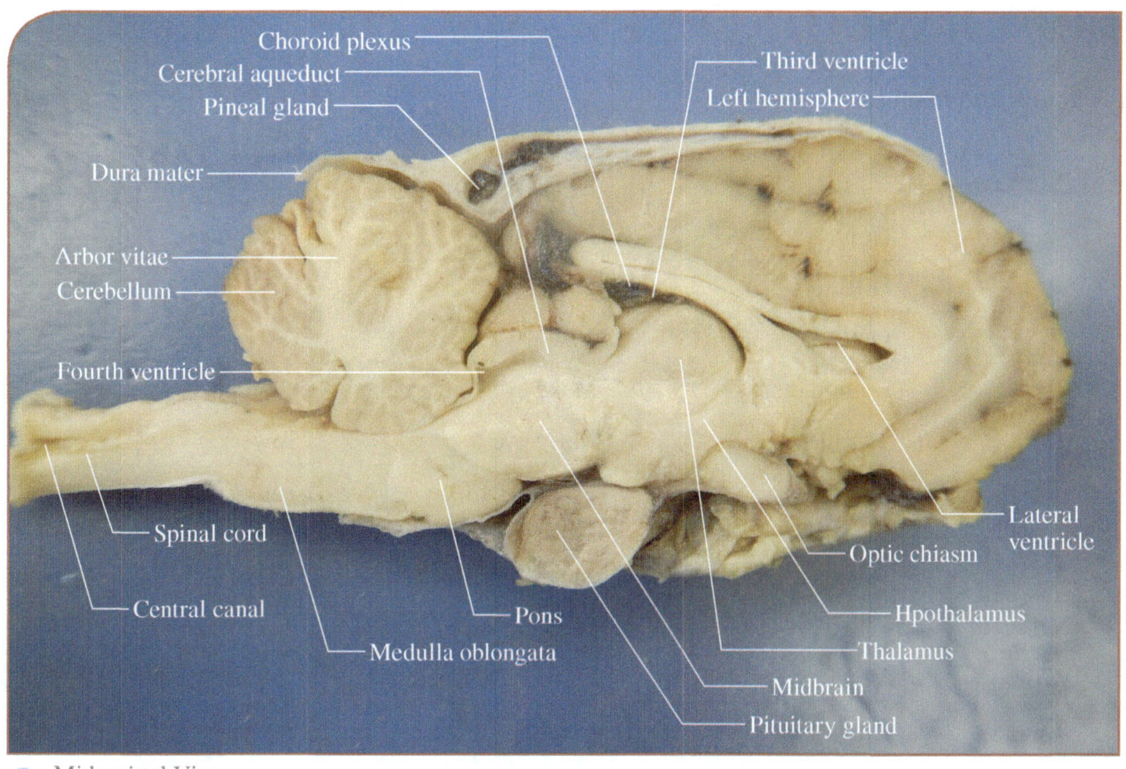

Midsagittal View

Whole Brain

1. Dura mater
2. Arachnoid/pia mater
3. Cerebellum
4. Cerebrum
5. Spinal cord
6. Temporal lobe
7. Frontal lobe
8. Parietal lobe
9. Occipital lobe
10. Right cerebral hemisphere
11. Longitudinal fissure
12. Pituitary gland (if still intact)
13. Pons
14. Medulla
15. Optic chiasm
16. Optic tract
17. Optic nerve
18. Olfactory bulb
19. Olfactory tract

Sagittal Cut Brain

1. Cerebral cortex
2. Cerebellum
3. Arbor vitae
4. Midbrain
5. Pons
6. Medulla oblongata
7. Corpus callosum
8. Sulcus
9. Gyrus
10. Thalamus
11. Hypothalamus
12. Cerebral aqueduct
13. 3rd ventricle
14. Choroid plexus
15. Spinal cord
16. 4th ventricle
17. Pineal gland
18. Optic chiasm

LIST OF CRANIAL NERVES

Know the names and numbers.

CN #	Name	Function	Mnemonic
I	Olfactory	Smell	On
II	Optic	Vision	Old
III	Oculomotor	Move eye/convergence	Olympus'
IV	Trochlear	Move eye	Towering
V	Trigeminal	Sensory of face mm. of mastication	Top
VI	Abducens	Move eye	A
VII	Facial	Facial muscles	Friendly
VIII	Vestibulocochlear	Hearing/balance	Viking
IX	Glossopharyngeal	Gag/say 'ah'	Grew
X	Vagus	Gag/say 'ah'	Vines
XI	Accessory	Shrug shoulders Rotate head	And
XII	Hypoglossal	Tongue movement	Hops

Cranial Nerves

L A B *10*

Cranial Nerves, Spinal Cord and Reflexes

Objectives

The purpose of this lab is to have the student know the structures and functions of the spinal cord and peripheral nerves. The student will also be able to identify the parts of a spinal reflex. The student will learn how to perform three different assessments including the cranial nerve tests, deep tendon reflexes, and dermatomal testing.

Prior to lab: Complete the Pre-lab.

Complete APR Dissection and Animations

During lab: Complete each objective prior to leaving lab.

- [ ] **Microscopy of Tissues**
    5th Cervical Vertebrae Model
    Spinal Cord Model
- [ ] **Cranial Nerve Assessment**
- [ ] **Deep Tendon Reflex Assessments**
    Patellar Tendon
    Achilles Tendon
    Triceps Tendon
    Biceps Tendon

- [ ] **Identify Components of a Reflex**
- [ ] **Perform Plantar (Babinski) Reflex**
- [ ] **Dermatomal Assessments**
- [ ] **Watch Nervous System Videos**
    Bell's Palsy
    Trigeminal Neuralgia
    Clonus

LAB 10 PRELAB—Cranial Nerves, Spinal Cord and Reflexes

Cranial Nerve Anatomy

Label the names of the cranial nerves next to the numbers.

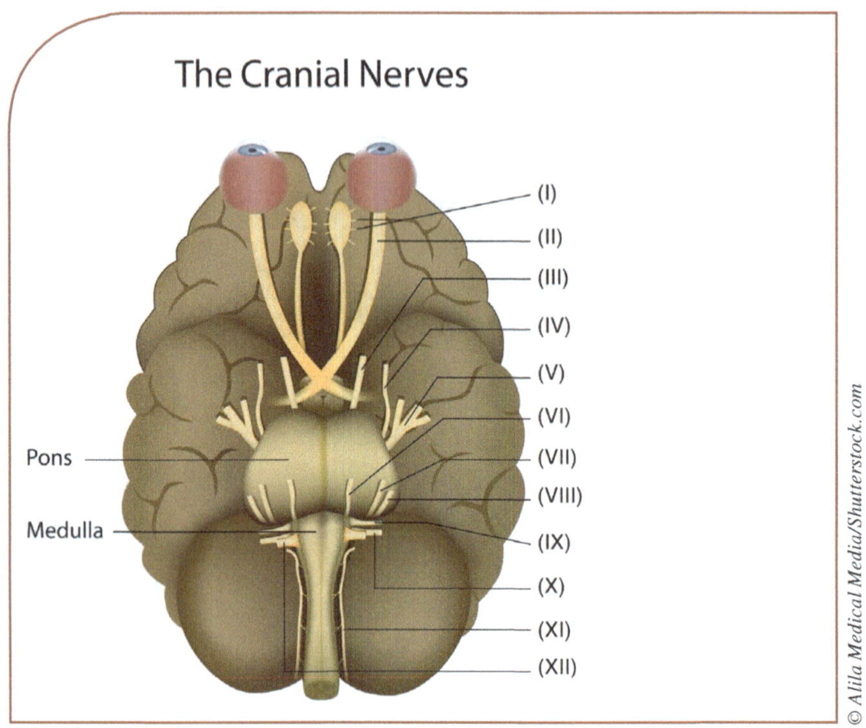

In the pictures below are 9 of the 12 cranial nerves. Label which cranial nerves are represented.

1. Olfactory
2. Optic
3. Oculomotor
4. Trochlear and abducens

5. Trigeminal
6. Facial
7. Vagus
8. Accessory

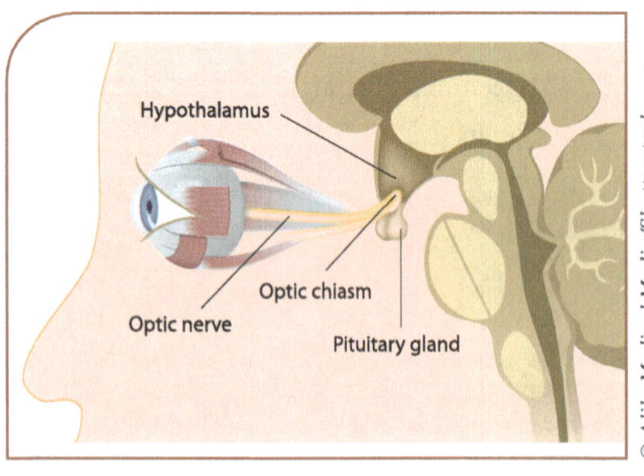

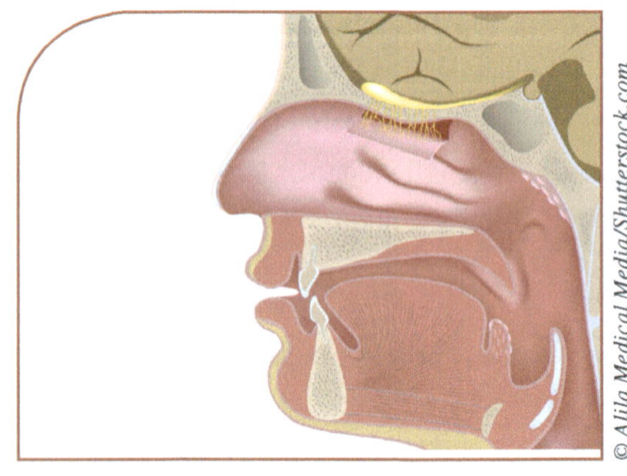

___________________________   ___________________________

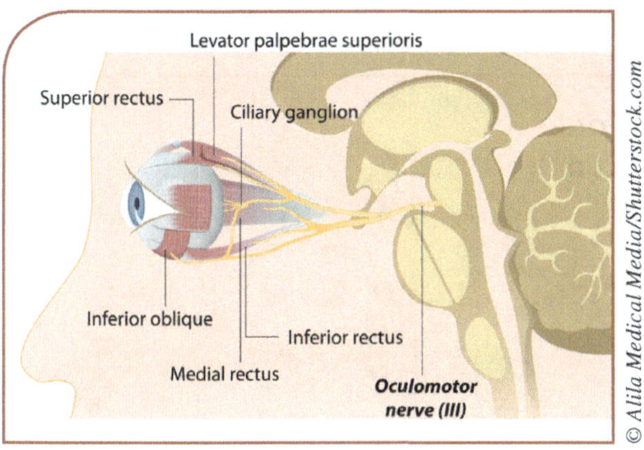

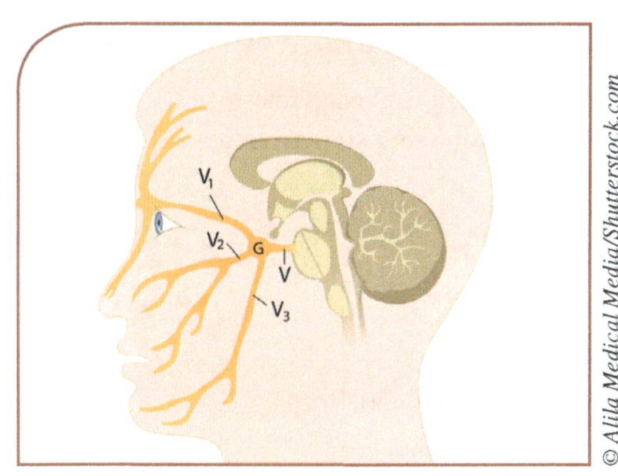

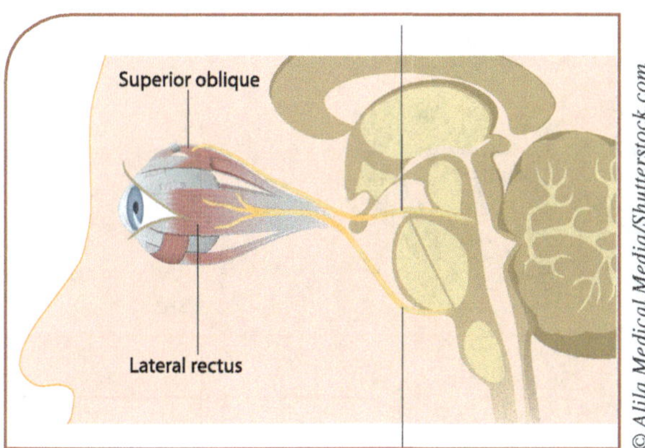

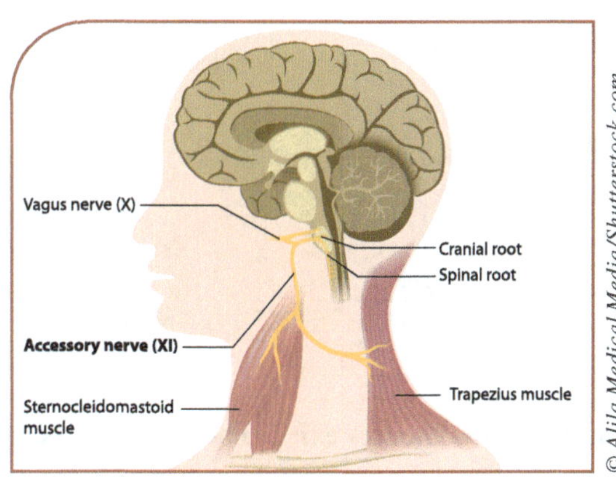

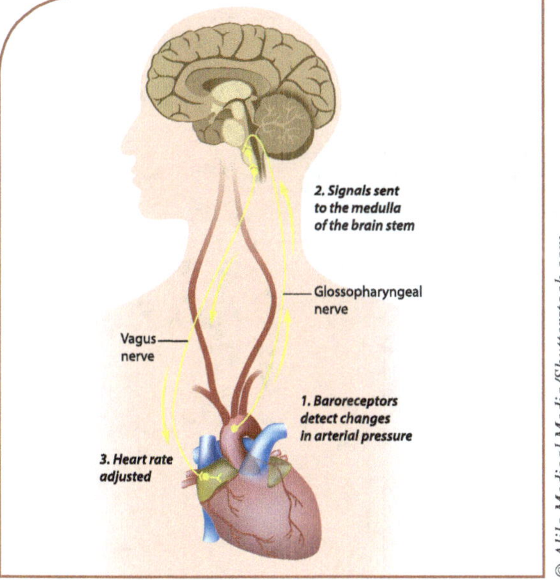

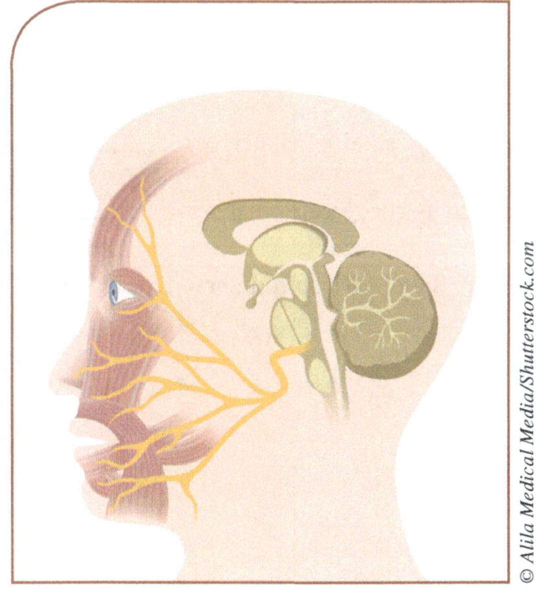

Which cranial nerves are missing? _______________________________

Spinal Cord and Spinal Nerve Anatomy

In the picture of the spinal cord cross section, label the following structures:

Central canal

Dorsal ramus

Dorsal root

Dorsal root ganglion

Ganglion of sympathetic trunk

Gray matter of spinal cord

Interneuron

Motor neuron

Rami communicans

Sensory neuron

Spinal nerve

Ventral ramus

Ventral root

White matter of spinal cord

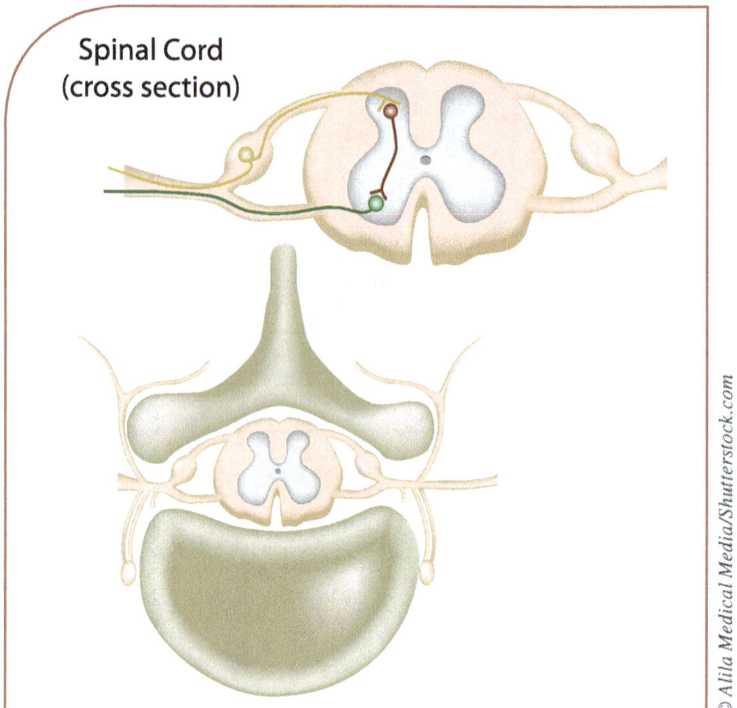

List how many spinal nerves exit in each level of the cord.

How many cervical spinal nerves? _______________________

How many thoracic spinal nerves? _______________________

How many lumbar spinal nerves? _______________________

How many sacral spinal nerves? _______________________

How many coccygeal spinal nerves? _______________________

In the drawing, label each spinal nerve from C1 to Co1.

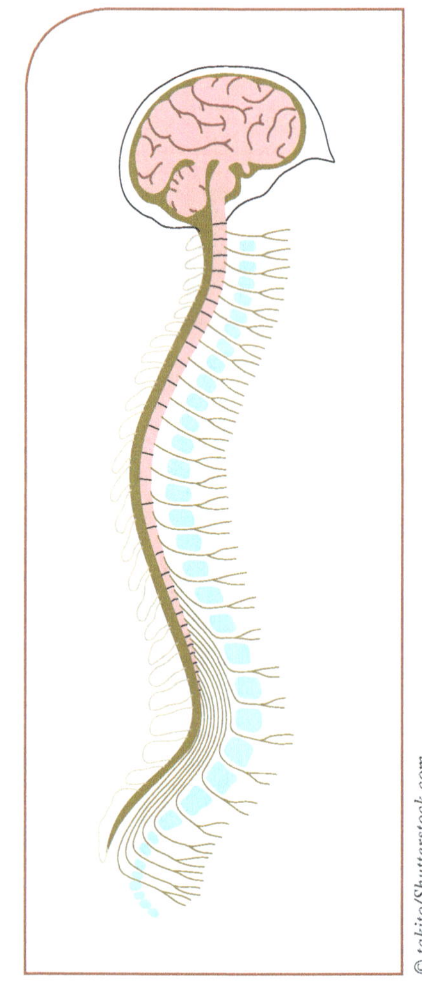

REFLEXES

We are going to look at reflexes over the next two labs. All reflexes have common characteristics. These include responses that are stereotyped, rapid, and involuntary. The pathway of reflexes called the reflex arc also has common basic structures that include a receptor, an afferent or sensory nerve fiber, an integrating center, an efferent or motor nerve fiber and an effector. In the drawing below, label the five components of a reflex.

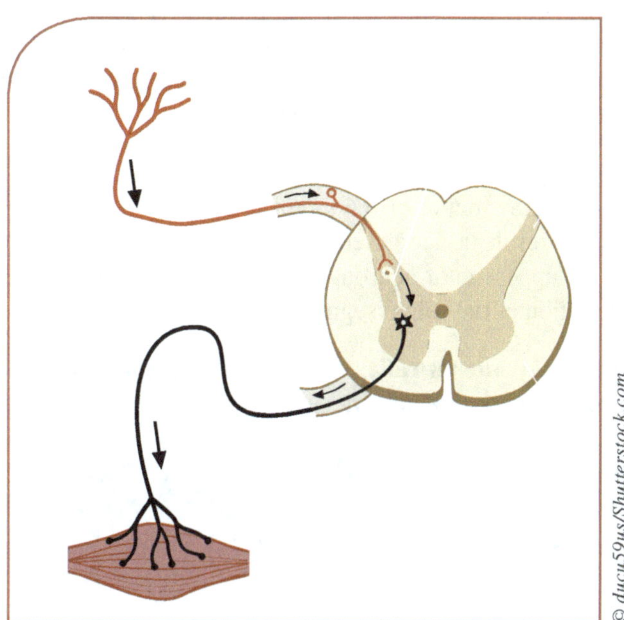

Reflexes are categorized as somatic or visceral (autonomic). Somatic reflexes are initiated by external stimuli, involve the somatic nervous system and skeletal muscle. Visceral reflexes are initiated by internal or external stimuli and involve the autonomic nervous system terminating on smooth muscle. Somatic reflexes include subcategories of reflexes. These include stretch reflexes or deep tendon reflexes, flexor withdrawal reflex, crossed extension reflex and primitive (neonatal) reflexes. Lab 11 will detail the visceral reflexes. In the chart below are some examples of somatic and visceral reflexes, check if the reflex is somatic or visceral.

Somatic and Visceral Reflexes		
Reflex	**Somatic**	**Visceral**
Gag reflex (CN IX and CN X)		
Achilles tendon reflex		
Flexor withdrawal reflex		
Triceps tendon reflex		
Cardiac baroreceptor reflex		
Babinski (plantar) reflex		
Pupillary light reflex (CN II)		
Rooting reflex		
Palmar reflex		
Micturition reflex		

Reflexes are one test used to examine a patient's neurological status. A common reflex test that we will do in lab is a stretch reflex (also called a deep tendon reflex) using the reflex hammer to quickly stretch a muscle. The reflex hammer hits the tendon which quickly stretches the muscle. The reflex arc is stimulated and elicits an involuntary contraction in that muscle in order to protect the muscle from being overstretched.

A stretch reflex is objectively assigned a grade from 0 to 4+ and documented in a patient chart. Normal is designated 2+. The designation 1+ means a sluggish or depressed reflex. Reflexes that are noticeably more brisk than usual are designated 3+, while 4+ means that the reflex is hyperactive and clonus is present. Clonus is a repetitive, rhythmic reflex response elicited by manually stretching the tendon. This clonus may be sustained as long as the tendon is manually stretched or may stop after up to a few beats. Almost any grade of reflex (outside of sustained clonus) can be normal. Asymmetry of reflexes is a key to determine pathology.

Decreased reflexes should lead to suspicion that the reflex arc has been affected. This could be the sensory nerve fiber but may also be the spinal cord gray matter or the motor fiber. This motor efferent fiber is termed the lower motor neuron (LMN). LMN lesions result in decreased reflexes. The descending motor tracts from the cerebral cortex and brain stem are called the upper motor neurons (UMN). Lesions of UMNs result in increased reflexes at the spinal cord by decreasing inhibition of the spinal segment.

Below are some pathologies that may alter a reflex. Put a check in the box that a patient would typically express based on what you know of the muscle and nervous system.

	Absent/Hyporeflexic	Normal	Hyperreflexic/Clonus
Cerebral stroke			
Damaged or cut spinal nerve (i.e. L4)			
Spinal cord tumor			
Parkinson's disease			
Polio			
Diabetic neuropathy			
Sciatica (peripheral nerve injury)			
Guillian-Barre			
ALS			
Traumatic brain injury to frontal lobe			
Traumatic spinal cord injury			

MODULE 7: NERVOUS SYSTEM

Dissection

Brain: Lateral

LAYER 1: Cutaneous distribution of mandibular nerve, cutaneous distribution of maxillary nerve, cutaneous distribution of ophthalmic nerve. (All these can be called as sensory branches of CNV or sensory branches of trigeminal nerve)

Cranial Nerves: CN I Olfactory

LAYER 3: Olfactory bulb, olfactory nerve (CN I), olfactory tract

Cranial Nerves: CN II Optic

LAYER 4: Optic nerve (CN II)

Cranial Nerves: CN V Trigeminal

LAYER 3: Trigeminal nerve (CN V)

Cranial Nerves: CN VII Facial

LAYER 3: Facial nerve (CN VII)

LAYER 4: Motor component of facial nerve, parasympathetic component of facial nerve.

What component of the facial innervates the muscles of facial expression? ___________________________

The __________________ component of CN VII goes to the lacrimal gland and the

__________________ glands.

Spinal Cord: Overview

LAYER 1: Know that each pin indicates a specific dermatome. You do not need to know each individual dermatome.

Define dermatome. ____

LAYER 3: Dura mater

LAYER 4: Cauda equina, dorsal root ganglion

The cauda equina are bundles of nerve roots below ____.

The dorsal root ganglion a collection of cell bodies from the   *sensory   motor*   neurons.   (circle)

LAYER 5: Cervical region of the spinal cord, filum terminale, lumbar region of spinal cord, sacral region of spinal cord, spinal cord, sacral region of spinal cord, thoracic region of spinal cord

What vertebral level does the spinal cord end? _______________________________________

The filum terminale is a nerve.   *TRUE   FALSE*   (circle)

Spinal Cord: Cervical Region

LAYER 4: Cervical region of spinal cord, dorsal ramus, dorsal root, dorsal root ganglion, spinal nerve, ventral ramus, ventral root

How many pairs of cervical spinal nerves are there? _______________________________________

Spinal Cord: Thoracic Region

LAYER 4: Thoracic region of spinal cord

How many pairs of thoracic spinal nerves are there? _______________________________________

Spinal Cord: Lumbar Region

LAYER 3: Dura mater

LAYER 4: Cauda equina, dorsal root ganglion, filumterminale, spinal nerve.

How many pairs of thoracic spinal nerves are there? _______________________________________

Spinal Cord: Typical Spinal Nerve

LAYER 1: Dorsal ramus, dorsal root, dorsal root ganglion, dura mater, epidural space, ganglion of sympathetic trunk, gray matter of spinal cord, gray ramus communicans (can be called ramus comminicans), spinal nerve, ventral ramus, ventral root, white matter of spinal cord, white ramus communicans (can be called ramus comminicans)

Circle what information each the nerve structure transmits.

Dorsal ramus	*Sensory*	*Motor*	*Both*
Dorsal root	*Sensory*	*Motor*	*Both*
Ramus communicans	*Sensory*	*Motor*	*Both*
Spinal nerve	*Sensory*	*Motor*	*Both*
Ventral ramus	*Sensory*	*Motor*	*Both*
Ventral root	*Sensory*	*Motor*	*Both*

The grey matter on the spinal cord is   *deep   superficial*   . (circle)

The rami communicans contains   *sympathetic   parasympathetic*   information.   (circle)

How many pairs of spinal nerves are there? ____

Peripheral Nerves: Cervical Plexus

LAYER 4: Phrenic nerve, sympathetic trunk

What muscle does the phrenic nerve innervate? ____

LAYER 5: Accessory nerve (CN XI), brachial plexus

What large muscles does the accessory nerve innerve? ____

Peripheral Nerves: Brachial Plexus

LAYER 5: Brachial plexus and branches, median nerve with spinal roots, musculocutaneous nerve with spinal nerve roots, ulnar nerve with spinal nerve roots

The brachial plexus composes the ventral rami from spinal nerves C5 to ________________.

LAYER 6: Axillary nerve with spinal nerve roots, radial nerve with spinal nerve roots

Peripheral Nerves: Upper Limb—Anterior

LAYER 3: Median nerve with spinal nerve roots, musculocutaneous nerve with spinal nerve roots, ulnar nerve with spinal nerve roots

Peripheral Nerves: Upper Limb—Posterior

LAYER 3: Axillary nerve with spinal nerve roots, radial nerve with spinal nerve roots

Peripheral Nerves: Hand—Palmar

LAYER 6: Brachial plexus and terminal branches, median nerve and branches, ulnar nerve and branches

Peripheral Nerves: Trunk

LAYER 1: T4 dermatome, T10 dermatome

Peripheral Nerves: Lower Limb—Anterior

LAYER 1: L2 dermatome, L3 dermatome, L4 dermatome, L5 dermatome, S1 dermatome

Which dermatome is over the skin of the lateral knee? ____

If a patient complained of numbness over his great toe and medial leg, what spinal nerve would you be suspect to be compromised? ____

LAYER 2: Femoral nerve and branches, lumbosacral plexus

Peripheral Nerves: Lower Limb—Posterior

LAYER 2: Sciatic nerve and branches

Peripheral Nerves: Hip and Thigh—Anterior

LAYER 5: Femoral nerve and branches

What peripheral nerve innervates the quadriceps? ____

Peripheral Nerves: Hip and Thigh—Posterior

LAYER 4: Sciatic nerve

Animation: Typical Spinal Nerve

There are ________________ pairs of sacral nerves and ________________ pair of coccygeal nerves.

The rootlets connect the spinal cord to the ________________. The ventral roots are the

________________ or efferent and the dorsal roots are the sensory or ________________.

When the dorsal and ventral roots unite, they form the ____.

What is a "mixed nerve"? ____

How does the spinal nerve exit the vertebral foramen? ____

What rami innervate the muscles of the limbs? ____

What rami innervate the muscles of the back? ____

Animation: Reflex Arc

Where is a reflex initiated? ____

After the receptor has been stimulated, an _________________ impulse travels along the axon of a

_________________ neuron.

An interneuron is only found in the central nervous system.   *TRUE   FALSE*   (circle)

An efferent impulse travels down a _________________ neuron to the _________________ organ.

An example of an effector is ____.

A reflex arc happens   *slowly   rapidly* . (circle)

SPINAL CORD MODELS

Be able to identify to following structures.

Grey matter	White matter	Ventral ramus
a. Dorsal horn	Anterior median sulcus	Dorsal ramus
b. Lateral horn	Posterior median sulcus	Rami communicans
c. Ventral horn	Dorsal (posterior) root	Sympathetic ganglion
	Spinal nerve	

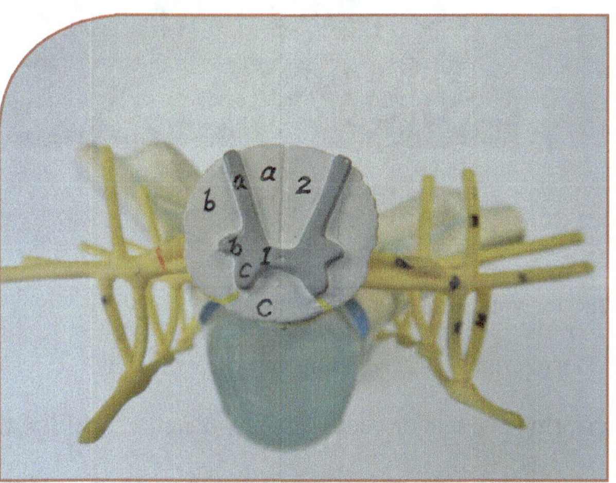

5th Cervical Vertebra Model

Body of 5th cervical vertebra

Grey matter

Ventral horn (anterior horn)

Dorsal horn (posterior horn)

Central canal

Ventral (anterior) root

Dorsal (posterior) root

Spinal nerve

Dorsal root ganglion (DRG)

Ventral ramus

Dorsal ramus

Rami communicans

Sympathetic trunk

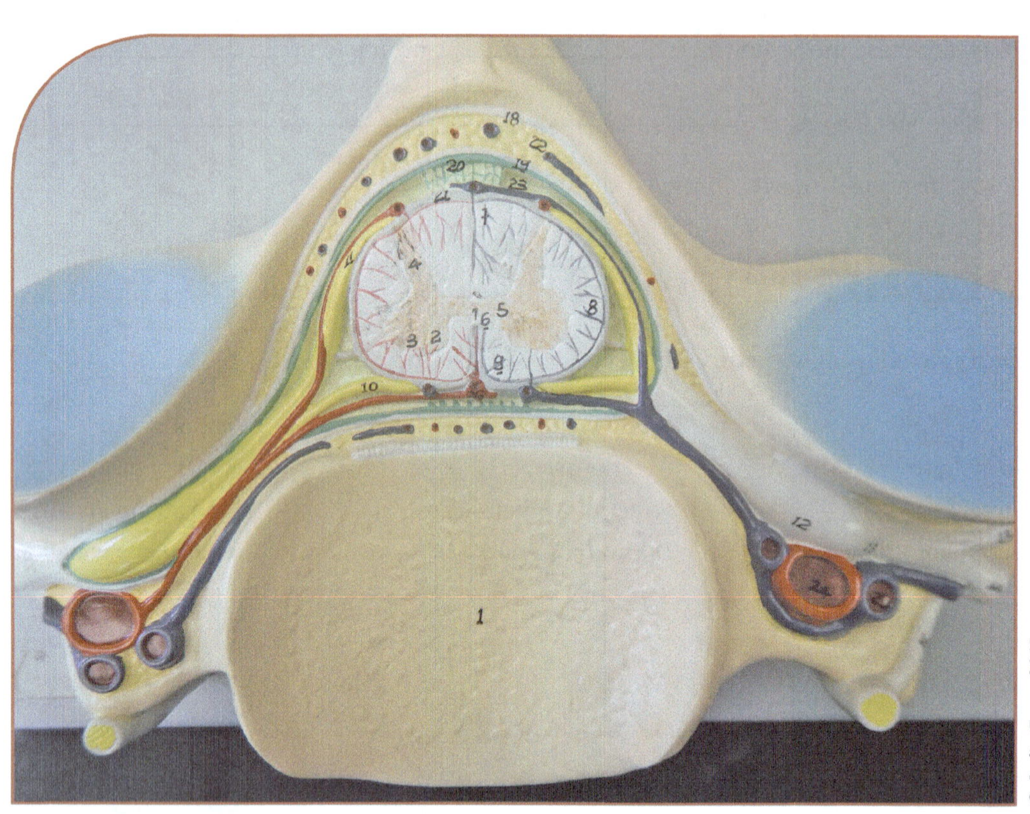

© Jodie Gerts, 2015

Neurological Assessments

Practice your approach to a patient.

Perform cranial nerve testing and assess results.

Perform deep tendon reflex testing—Patellar, achilles, triceps, biceps and assess results.

Perform upper extremity dermatomal assessment and interpret results.

Approach to Patient

1. Give your patient your full attention.
2. Acknowledge the patient by the name they prefer to be called. Introduce yourself to your patient.
3. Make eye contact with the patient.
4. Be aware of your body language and its subconscious meaning.
5. Explain what you are doing at each step and the reasons for doing it.
6. Show the patient the tools you will be using, if appropriate.
7. If appropriate, demonstrate to patient what test you will administer with their eyes open and on the unaffected side of the body so they know the appropriate response.
8. Record the response.
9. Interpret the results.
10. Thank the patient for their cooperation. And ask if they have any questions. Maintain open communication throughout examination to ensure patient trust and compliance.

THE CRANIAL NERVE ASSESSMENT

Listed are the primary tests of the cranial nerves.

Primary Tests of the Cranial Nerves			
CN #	Sensory/ Motor Testing	Test	What You Might Observe
I	Sensory	Ask the pt to **SMELL SOMETHING** testing each nostril separately (without looking), e.g., an alcohol prep pad, food items.	The patient may complain about the lack of odors associated with their meals.
II	Sensory	**FIELD OF VISION:** Face the pt nose-to-nose having them look you directly in the eyes. Wiggle your finger at the edges of your field of vision and ask the pt when they can see motion. **ACUITY:** Unilaterally test each eye using the Snellen eye chart or Rosenbaum chart. If using Snellen, have the patient stand 20 feet away and ask them to read the smallest line they can see. If using Rosenbaum chart, hold chart 14 inches from eyes. Have patient read the chart. **PERRL:** "pupils equally round and reactive to light". Cover each eye at a time and shine a bright light directly onto the uncovered eye. Both pupils should respond.	The pt may complain of missing parts of their field of vision. They may be unable to read or see distant objects. Constricted pupils.

Primary Tests of the Cranial Nerves *(continued)*

CN #	Sensory/ Motor Testing	Test	What You Might Observe
III, IV, VI	Motor	**MOVE THE EYE:** Have the pt hold their head still and follow your finger through all six cardinal fields of vision. Both eyes should track the movement of your finger. **CONVERGENCE:** Have the patient follow your finger as you move it towards their nose.	You may note that both eyes do not seem to track your movement. Nystagmus may be seen.
V	Sensory and motor	<u>SENSORY</u>: CN V controls the **SENSATION OF THE FACE**. Light touch, dull, or sharp (i.e. cotton ball) over the forehead, the cheeks, and along the jaw. It should feel the same on both sides. You could also test pain and temperature in the same manner. <u>MOTOR</u>: Have pt. **clench their jaws** and palpate the firmness of the right and left masseter and temporalis mm.	The patient may have trouble chewing their food. They may complain of pain or loss of sensation in any of the three cutaneous distributions on the face.
VII	Sensory and motor (not tested)	Ask the pt to **MOVE FACIAL MUSCLES:** frown, wrinkle your forehead; close your eyes tightly, smile, purse your lips, puff out cheeks.	You may note an asymmetry in any of the patient's facial expressions.
VIII	Sensory	**HEARING:** Have patient occlude one ear. Stand one foot away from patient and whisper word into patient's ear and have them repeat it. **EQUILIBRIUM:** Have the patient stand on one leg with eyes closed. Test bilaterally.	The pt may have difficulty hearing whispered words. Note if you must raise voice or stand closer. The pt may be unable to balance without visual input.
IX, X	Sensory and motor (not tested)	<u>MOTOR</u>: Have the **PT SAY "AH."** The uvula and soft palate should have even movement. Assess **gag reflex** using tongue blade. Stroke the palatial arch using blade. The patient should gag and elevate arch.	The pt may have difficulty swallowing. The pt's voice may be hoarse (or has changed). Deviation of uvula to unaffected side.
XI	Motor	Innervates the sternocleidomastoid and trapezius mm. Have patient shrug their shoulders then apply resistance to test the trapezius. Have the patient turn head side to side to test the sternocleidomastoid.	Inequality between right and left sides.
XII	Motor	Have the pt **STICK OUT THEIR TONGUE** and move it side to side.	Inequality between right and left sides.

THE DEEP TENDON REFLEX ASSESSMENT

Procedures

Patellar Reflex: L4 Spinal Nerve Root

1. Have your partner sit on the edge of a table with his/her legs dangling loosing over the edge.

2. Clinician applies the stimulus by tapping, firmly, but carefully, over the patellar tendon with the reflex hammer.

3. Test both sides. Record your results.

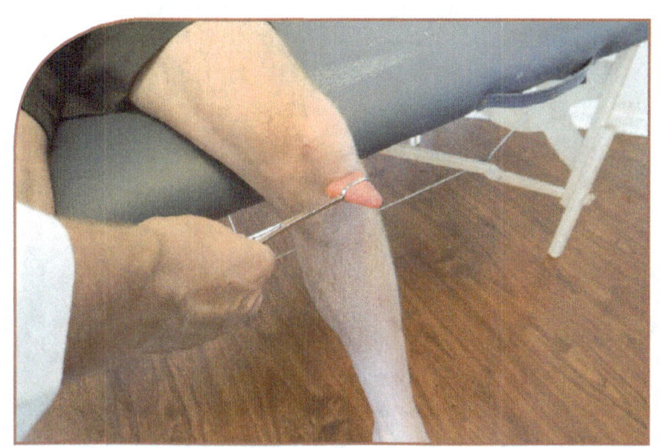

© Jodie Gerts, 2015

Achilles Reflex: S1 Spinal Nerve Root

1. Remove the shoe and slightly dorsiflex the foot to increase the tension in the *gastrocnemius and soleus* muscle. The clinician should support the foot in the hand.

2. Sharply tap the Achilles tendon above the calcaneous with the reflex hammer.

3. Test both sides. Record your results.

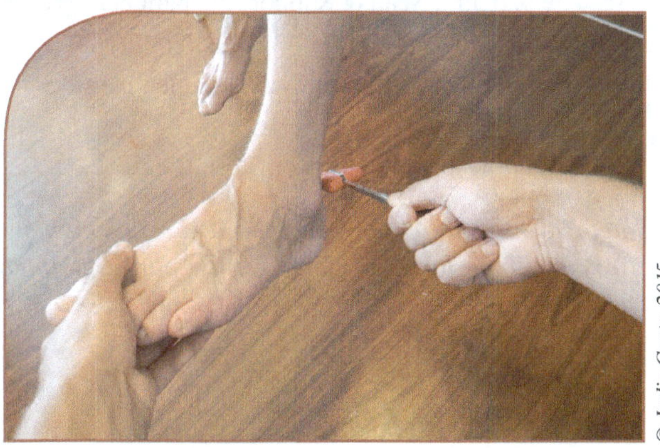

© Jodie Gerts, 2015

Triceps Tendon Reflex: C7 Spinal Nerve Root

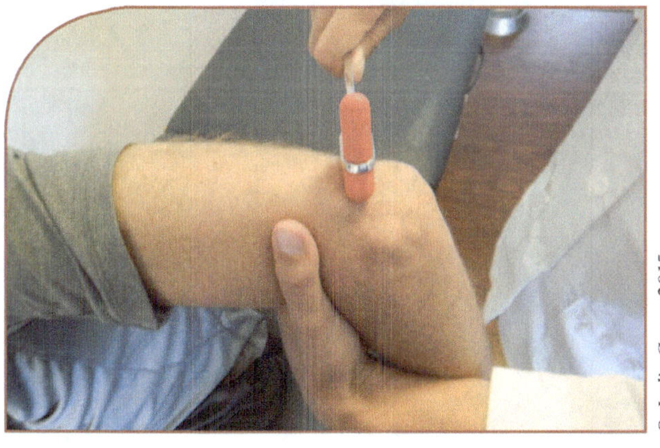

© Jodie Gerts, 2015

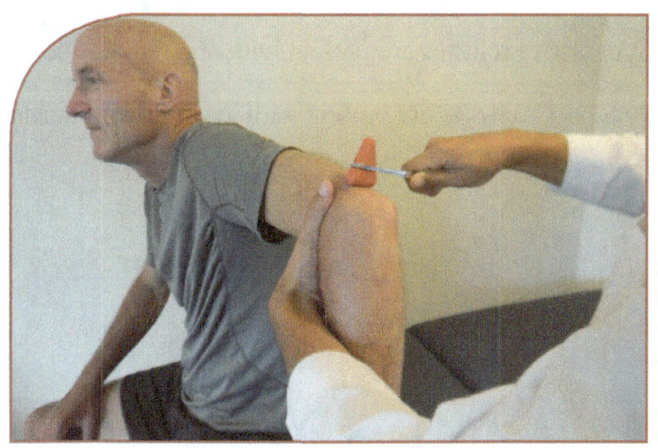

© Jodie Gerts, 2015

1. The patient is seated. Hold the patient's humerus distally while the arm is in 90 degrees of abduction and full internal rotation. The forearm is hanging freely.

2. Sharply tap the triceps tendon to elicit the reflex.

3. Test both sides. Record your results.

Biceps Tendon Reflex: C5 And C6 Spinal Nerve Root

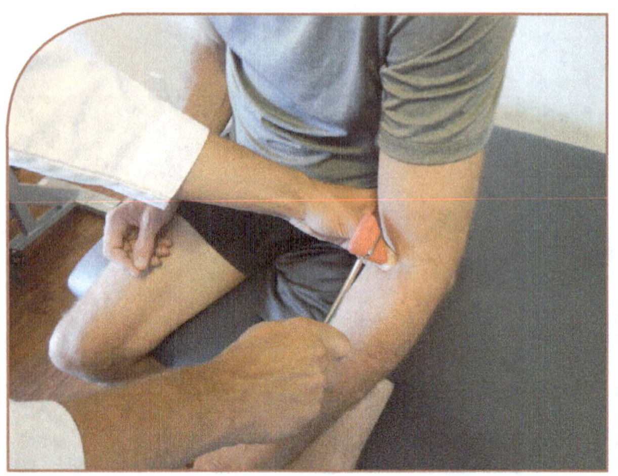
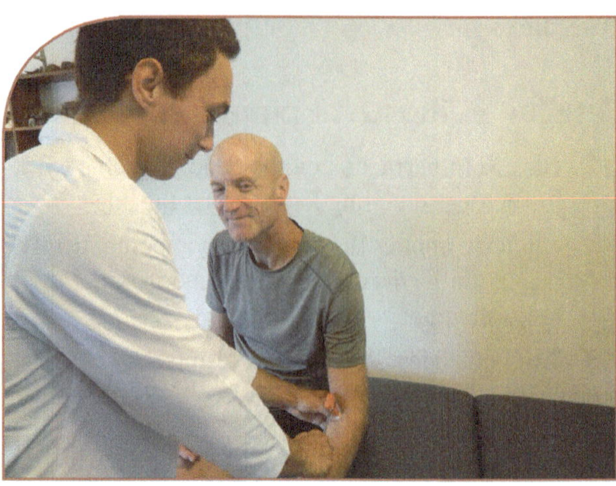

1. The patient is seated. Support the patient's forearm with your arm while placing your thumb over the patient's biceps tendon and applying pressure to the tendon.

2. Sharply hit your thumb with the reflex hammer to elicit the contraction.

3. Test both sides. Record your results. ____

Note: Have the patient perform the Jendrassik maneuver if you are having difficulty eliciting reflex on the lower extremity. To perform the maneuver ask patient to interlock fingers of both hands and pull apart.

- Reflex grading:
    - 0: *Complete absence*
    - 1: *Diminished*
    - 2: *Normal reflex*
    - 3: *Hyperactive reflex*
    - 4: *Clonus present*

Hyperactive reflexes are present with upper motor neuron lesions.

Decreased reflexes are present with lower motor neuron lesions.

Monosynaptic Somatic Reflex

Know the following terms and be able to name the parts on a model and the excitatory or inhibitory signals and the direction of the signals.

Sensory receptor

Sensory (afferent) neuron including dorsal root ganglion, dorsal root, spinal nerve and dorsal horn on spinal cord

Integrating center including dorsal horn, interneuron, ventral horn

Motor (efferent) neuron including ventral root and spinal nerve

Effector including name of the skeletal muscle

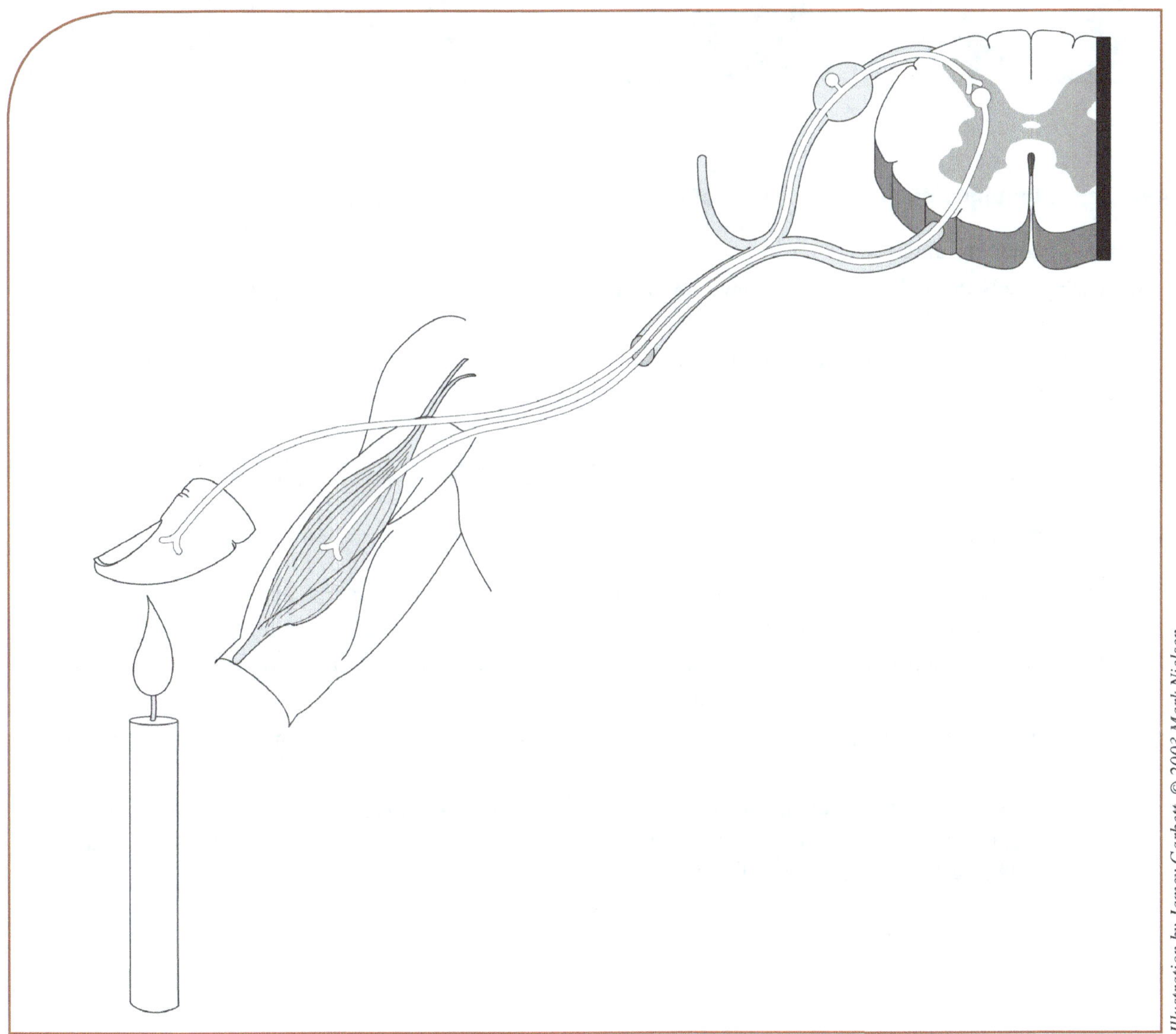

Plantar (Babinski) Reflex

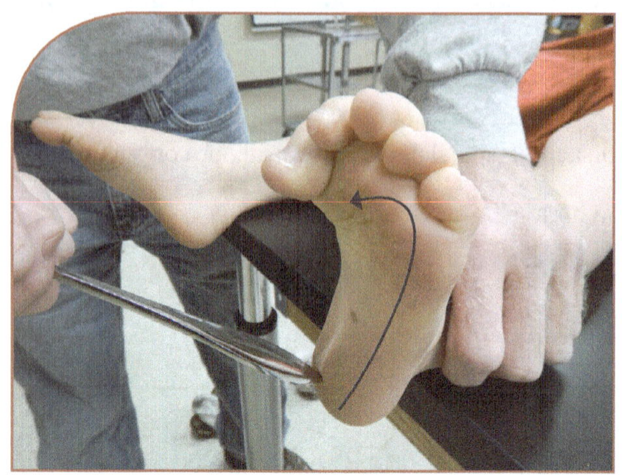

Normal Plantar Response

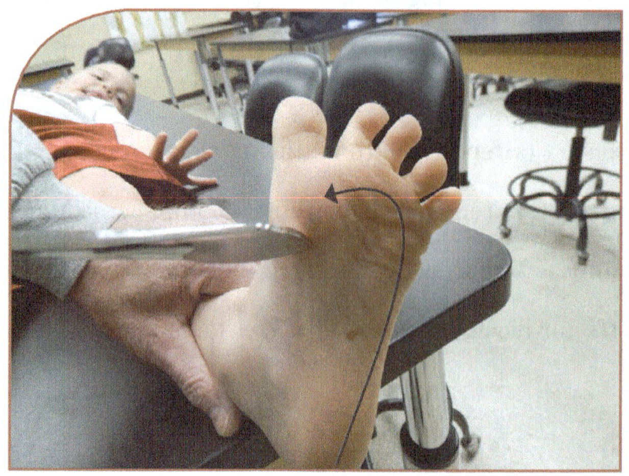

Babinski Sign

Assessment for upper motor neuron lesion of pyramidal tract (corticospinal tract)

1. Have patient in supine position with shoe and sock off foot.
2. Briskly apply the end of a reflex hammer on the plantar surface of foot from the heel through the metatarsal heads.
3. Normal reflex is flexion of the toes. Abnormal finding is extension of the great toe and abduction of the lesser toes.

UPPER EXTREMITY DERMATOMAL ASSESSMENT

Assessment Technique

1. The patient should be in a comfortable position that allows easy access to the required areas of the skin. Dermatome testing should not occur through clothing.
2. Prior to testing the patient should be educated about the purpose of the test. (I am going to test your ability to sense my touch at several places on your arms/legs).
3. The patient's eyes should be closed for the duration of the test to eliminate visual input that could skew the results.
4. The touch should be light enough that contact with the epidermis is made but the skin is not indented. A clinician can use a finger or a wisp of cotton. The right and left sides should be tested to allow the patient to compare an affected to an unaffected side.

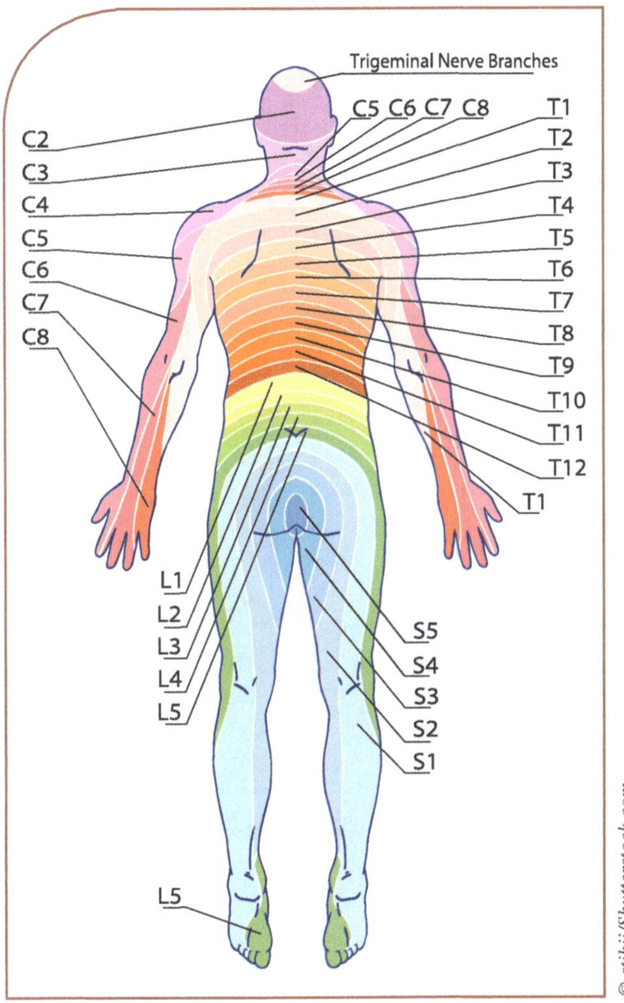

Areas

C4: Acromion

C5: Lateral brachium

C6: Thumb

C7: Middle finger

C8: Pinky

T1: Medial antecubital region

T4: Chest at level of nipple (will not test in class)

T10: Abdomen at level of umbilicus (will not test in class)

VIDEOS

Bell's Palsy

Trigeminal Neuralgia

Clonus

L A B *11*

Autonomic Nervous System and General and Special Senses

Objectives

The purpose of this lab is to have the student know the structures and functions of the eye and ear. The student will assess general sensation using Semmes-Weinstein monofilament testing. Additionally, the student will learn the anatomy of the autonomic nervous system and be able to analyze consequences of pharmacologic medications that act on the receptors of the autonomic nervous system.

Prior to lab: Complete the Pre-lab.

Complete APR Dissection, Histology, Imaging and Animations

During lab: Complete each objective prior to leaving lab.

☐ **Dissection of Cows Eye**

☐ **Models**
   Ear
   Eye

☐ **Semmes-Weinstein Monofilament Testing**

☐ **Worksheet on the Autonomic Nervous System**

LAB 11 PRELAB—Autonomic Nervous System and General and Special Senses

Identify and color the following anatomical structures of the ear:

Auricle

Auditory tube or
eustacian tube

Auditory canal

Cochlea

Inner ear

Incus

Malleus

Middle ear

Outer ear

Oval window

Round window

Semicircular ducts

Stapes

Petrous part of temporal bone

Tympanic membrane

Vestibulocochlear nerve
(CN VIII)

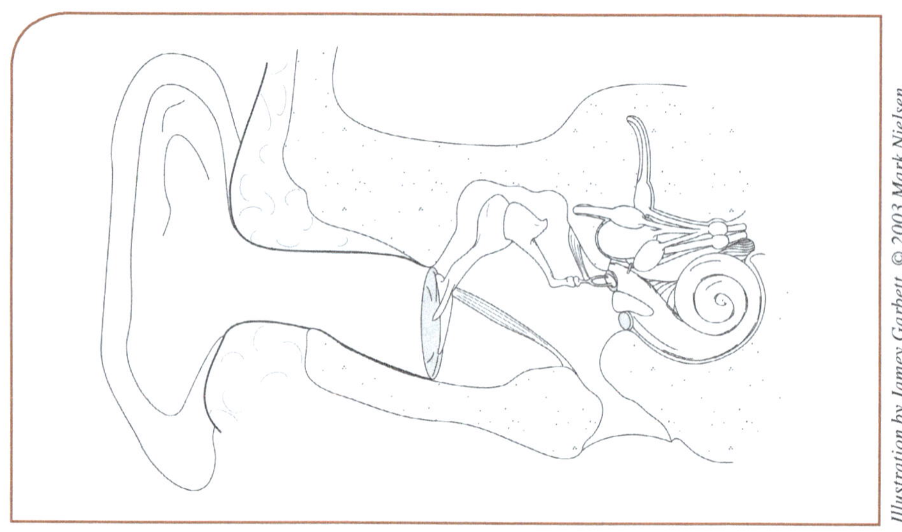

Identify the following anatomical structures of the inner ear:

Ampulla

Cochlea

Oval window

Saccule

Semicircular ducts

Utricle

Vestibulocochlear nerve
(CN VIII)

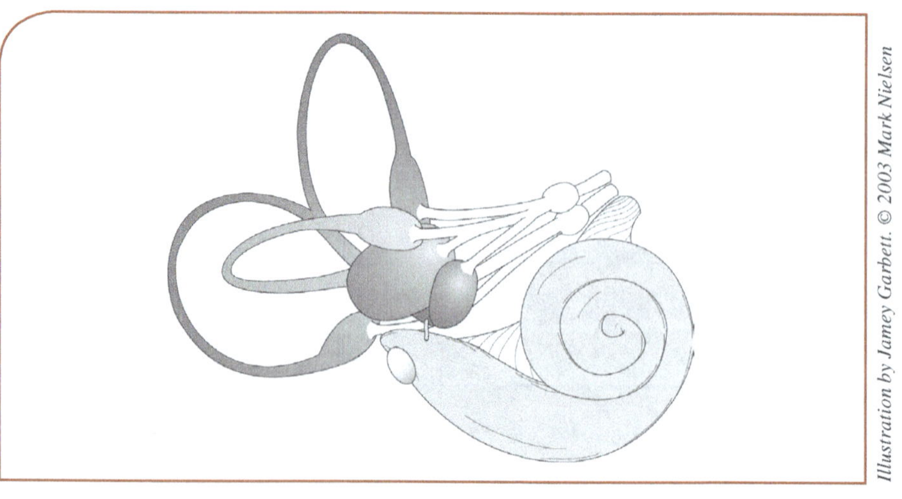

Identify and color the following structures of the cochlea.

Basilar membrane

Cochlear duct

Endolymph

Perilymph

Spiral organ

Scalivestibuli

Scali tympani

Vestibular membrane

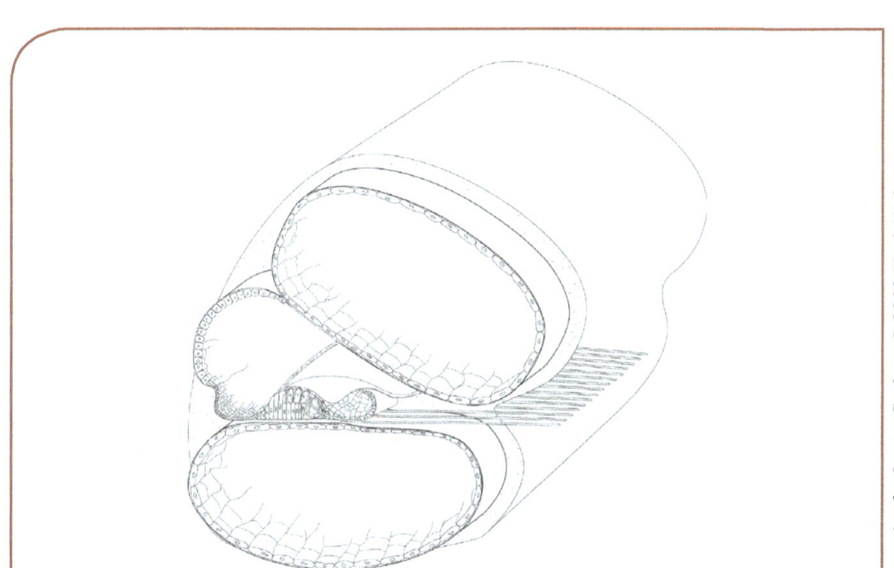

Identify and color the following structures of the spiral organ.

Basilar membrane

Cochlear nerve (CN VIII)
fibers

Hair cells

Tectorial membrane

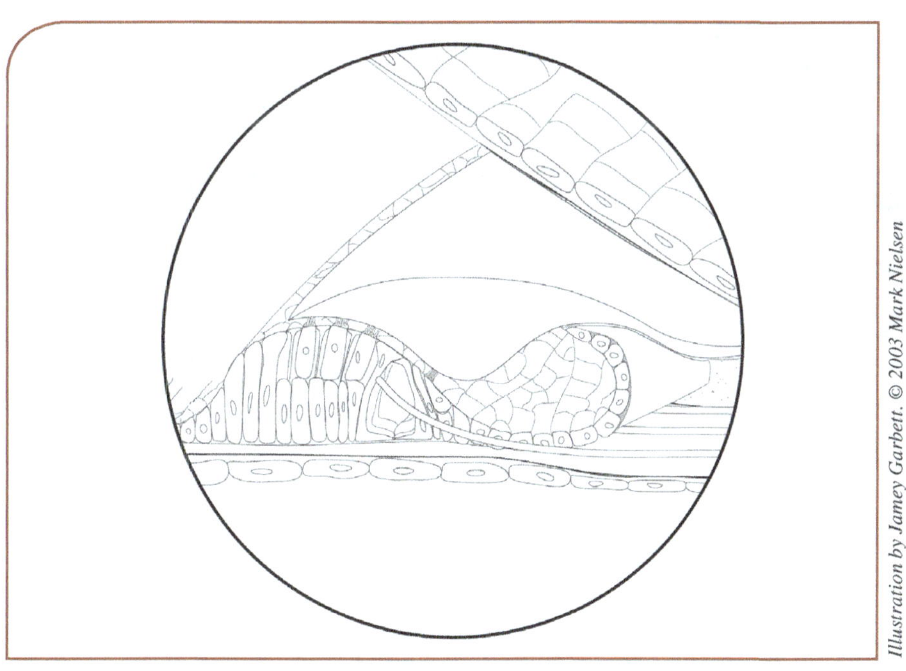

Identify and color the following anatomical structures of the eye.

Anterior cavity	Extraocular muscle	Ora serrata
Aqueous humor	Iris	Posterior cavity
Cornea	Lens	Retina
Choroid	Macula lutea with fovea centralis	Sclera
Ciliary body		Suspensory ligaments
Conjunctiva	Optic disk	Vitreous humor
	Optic nerve	

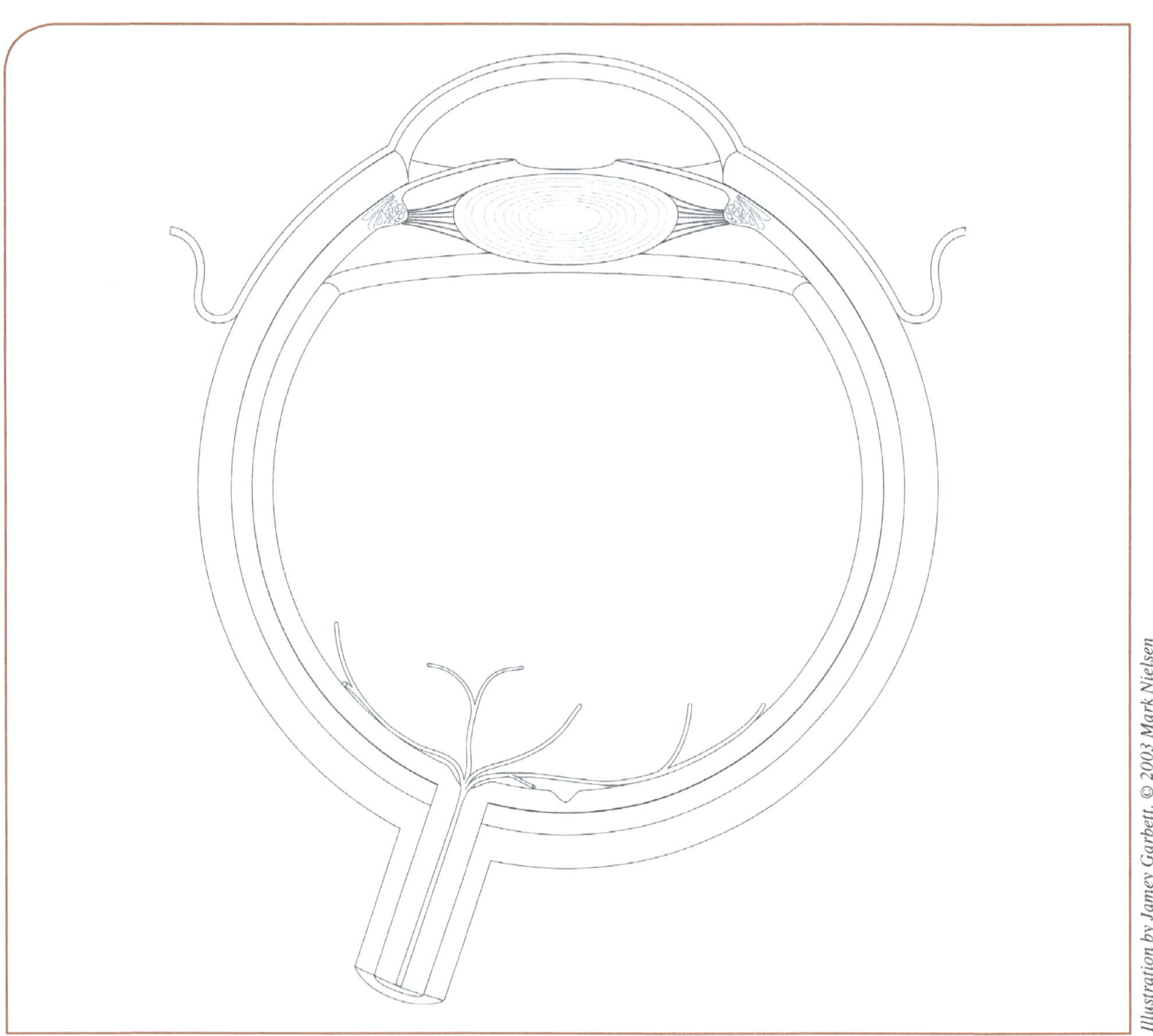

Shown in the illustration is the accommodation of the lens for near-sighted and far-sighted vision. Identify which picture shows the ciliary muscle relaxed and contracted. Identify which picture shows the suspensory ligament tense and relaxed. Identify which picture accommodates for near vision and distant vision.

Ciliary muscle: _______________________

Suspensory ligament: _______________________

Vision: _______________________

Ciliary muscle: _______________________

Suspensory ligament: _______________________

Vision: _______________________

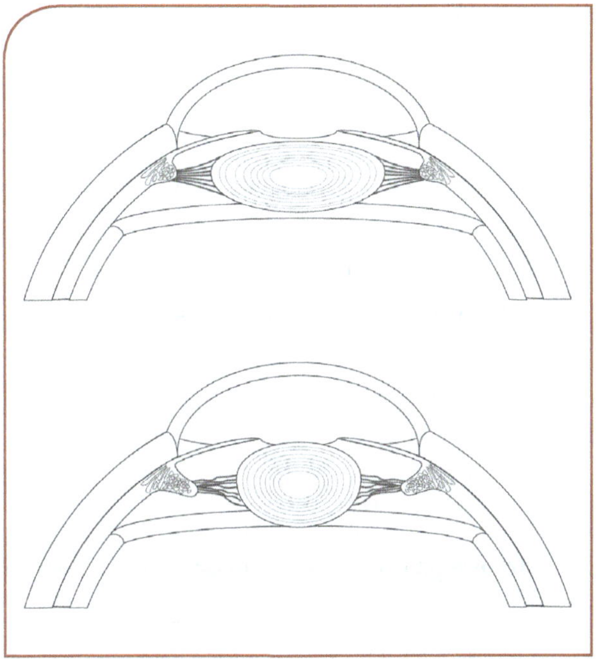

Shown in the images below are reflex pathways of the parasympathetic and sympathetic nervous systems on the descending colon and rectum. Identify the structures in each image and answer the following questions about each system.

Parasympathetic Structures

Dorsal root

Dorsal root ganglion (DRG)

Effector (colon)

Postganglionic parasympathetic fiber

Preganglionic sympathetic fiber

Sacral spinal cord

Sensory neuron from colon receptors

Terminal ganglia

Ventral root

Symapthetic Structures

Dorsal root

Dorsal root ganglion (DRG)

Effector (colon)

Postganglionic sympathetic fiber

Preganglionic parasympathetic fiber

Lumbar spinal cord

Sensory neuron from colon receptors

Ganglion (Inferior sympathetic)

Ventral root

Sympathetic trunk

White ramus communicans

Draw arrows on each image representing the direction of the nervous impulse.

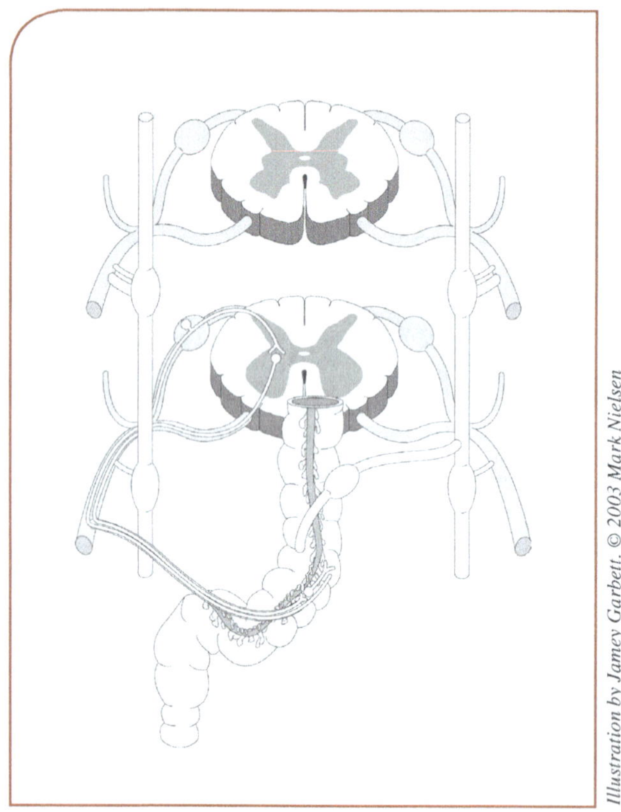

Parasympathetic Reflex

Illustration by Jamey Garbett. © 2003 Mark Nielsen

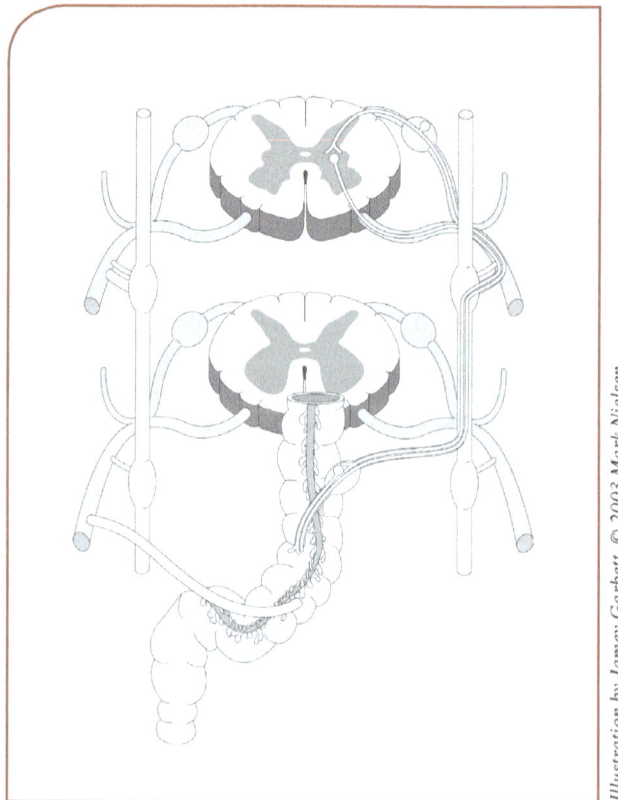
Sympathetic Reflex

Illustration by Jamey Garbett. © 2003 Mark Nielsen

MODULE 7: NERVOUS SYSTEM

Dissection

Sympathetic (ANS): Overview

LAYER 1: Sympathetic innervation of all effectors, sympathetic innervations of bronchus, sympathetic innervation of cardiac muscles and blood vessels, sympathetic innervation of pupillary dilator mm., sympathetic innervations of skin, sympathetic innervation of suprarenal medulla, sympathetic trunk, thoracolumbar (sympathetic) region of spinal cord, white ramus communicans

What color are the preganglionic fibers? ____

What color are the postganglionic fibers? ____

Which nerve fibers are shorter in the sympathetic system?   *Preganglionic   Postganglionic*   (circle)

When the sympathetic system fires, the bronchial tree   *dilates   constricts*   . (circle)

When the sympathetic system fires, the heart rate   *increases   decreases*   . (circle)

When the sympathetic system fires, the _______________ muscle in the skin contracts and the secretion of the sweat glands   *increase   decrease*   . (circle)

Firing of the sympathetic nerves to the suprarenal (adrenal) medulla stimulates the release of

_______________.

Conduction of the sympathetic impulse to the suprarenal (adrenal) medulla involves a *long   short*   preganglionic neuron and the   *long   short*   postganglionic neuron.   (circle)

The preganglionic sympathetic cell bodies are found in the   *thoracolumbar   craniosacral*   part of the CNS.   (circle)

Sympathetic (ANS): Thoracic Region

LAYER 4: Ganglion of sympathetic trunk, sympathetic trunk

Parasympathetic (ANS): Overview

LAYER 1: Parasympathetic component of facial nerve (CN VII), parasympathetic component of glossopharyngeal nerve (CN IX), parasympathetic component of vagus nerve (CN X), parasympathetic innervation of all effectors, parasympathetic innervation of cardiac muscle and vessels, parasympathetic innervation of bronchus, parasympathetic innervation of pupillary sphincter muscle, parasympathetic innervation of small intestine

What color are the preganglionic fibers? ____

What color are the postganglionic fibers? ____

Which nerve fibers are shorter in the parasympathetic system?   *Preganglionic   Postganglionic*   (circle)

What 4 cranial nerves are part of the parasympathetic nervous system? _______________________

When the parasympathetic system fires, the bronchial tree   *dilates   constricts*   . (circle)

When the parasympathetic system fires, the heart rate   *increases   decreases*   . (circle)

The parasympathetic nervous system sends fibers to the skin.   *True   False*   (circle)

When the parasympathetic system fires, the peristalsis in the GI system   *increases   decreases*   . (circle)

Conduction of the sympathetic impulse to the suprarenal (adrenal) medulla involves
a   *long   short*   preganglionic neuron and the   *long   short*   postganglionic neuron.   (circle)

The preganglionic sympathetic cell bodies are found in the   *thoracolumbar   craniosacral*   part of the CNS.   (circle)

Taste: Tongue—Superior

LAYER 2: Dorsum of tongue, lingual tonsil, special sensory of glossopharyngeal nerve, sensory of chordae tympani (can be called the sensory of CN VII), vallete papilla

What two cranial nerves are responsible for taste? _____________________________________

Cranial nerve ____________________ is responsible for the posterior 1/3rd of the taste.

The vallete papillae are located in the posterior 1/3rd of the tongue and contain numerous

____.

Smell: Inferior Brain

LAYER 1: Olfactory bulb, olfactory tract

Smell: Nasal Cavity—Lateral

LAYER 3: Olfactory bulb, olfactory nerve (CN I), olfactory tract

The olfactory nerves pass through the ___________________ of the ethmoid bone.

Hearing/Balance: Ear—Anterior

LAYER 1: Auricle of ear, cochlea, semicircular ducts, temporal lobe, tympanic membrane, utricle of ear, vestibulocochlear nerve (CN VIII)

What is the organ of equilibrium? ____

What is the organ of hearing? ____

What part of the temporal bone is the cochlea located? _______________________________

Define vestibular. ____

Vision: Inferior Brain

LAYER 1: Optic chiasm, optic nerve (CN II), optic tract

Vision: Eye—Lateral

LAYER 1: Iris, pupil

What structure allows light to enter the eye? _______________________________

LAYER 2: Anterior cavity, choroid, ciliary body, cornea, lens, optic disk, optic nerve (CN II), ora serrata, posterior cavity, retina, sclera, suspensory ligaments of lens

Match the structure to the function

______ Retina	1. Connects ciliary muscle to lens
______ Suspensory ligament	2. Space occupied by vitreous humor
______ Choroid	3. Responsible for special sensation of vision
______ Ciliary body	4. Anterior margin of retina connected to ciliary body
______ Optic disk	5. Protects and maintains shape of eye
______ Cornea	6. Location of photoreceptors
______ Lens	7. Space filled with aqueous humor
______ Sclera	8. Vascular layer to retina and absorption of light
______ Anterior cavity	9. Junction of optic nerve to retina and location of blind spot
______ Posterior cavity	10. Transparent connective tissue for light refraction and protection
______ Ora serrata	11. Muscle adjusts thickness of lens and produces aqueous humor
______ Optic nerve	12. Focuses light on retina

Vision: Orbit—Lateral

LAYER 5: Cornea, lens, optic nerve (CN II), pupil, sclera, suspensory ligaments of lens

Histology: Hearing/Balance—Cochlea: Low Magnification

Basilar membrane, cochlear duct, spiral organ, scala tympani, scala vestibuli, vestibular membrane

Histology: Hearing/Balance—Cochlea: High Magnification

Basilar membrane, Cochlear duct, spiral organ, scala tympani, scala vestibuli, tectorial membrane, vestibular membrane

Perilymph is found in the   *Scala vestibuli   Scala tympani   Cochlear duct*   (circle all that apply)

What membrane separates the scala tympani from the cochlear duct? ________________________

What membrane separates the scala vestibuli from the cochlear duct? ________________________

Histology: Hearing/Balance—Spiral Organ

Basilar membrane, outer hair cell (can be called hair cell), tectorial membrane

Histology: Taste—Vallate Papilla

Taste bud, vallate papilla

What are the five primary taste sensations? ________________________

What is the lifespan of a taste bud? ________________________

Imaging: Tympanic Membrane

Tympanic membrane, wall of external acoustic meatus

Imaging: Tympanic Membrane with Otitis Media

Tympanic membrane, wall of external acoustic meatus

Imaging: Retina

Fovea centralis, optic disk

What part of the retina in the macula lutea has the greatest visual acuity? ________________________

What photoreceptor cells are responsible for color vision?   *Rods   Cones*   (circle)

What part of the retina lacks photoreceptors? ________________________

Animation: Hearing

Sound waves enter the outer ear and strike the ___________________ causing vibration.

The mechanical vibration of the stapes deforms the ___________________.

That vibration is transmitted through the perilymph in the ___________________ causing the

___________________ to vibrate.

What type of sounds cause the basilar membrane to vibrate away from the oval window?

____

The movement of the basilar membrane is detected by ___________________ which transmit that information to the vestibulocochlear nerve.

When the vibrations reach the fluid of the scala tympani they are absorbed by the ___________________.

PART 1:

DISSECTION OF COWS EYE

Procedure

1. Rinse the eye with water.
2. On the outside of the eye, locate the following parts:
   - **Fat**—surrounds the eye and cushions it from shock
   - **Optic nerve**—a white cord on the back of the eye about 3 mm thick
   - **Extraocular muscles (EOM)**—reddish, flat muscles found around the eye to raise, lower, and rotate the eye
   - **Cornea**—a clear covering over the front of the eye (preservative often makes this appear cloudy)

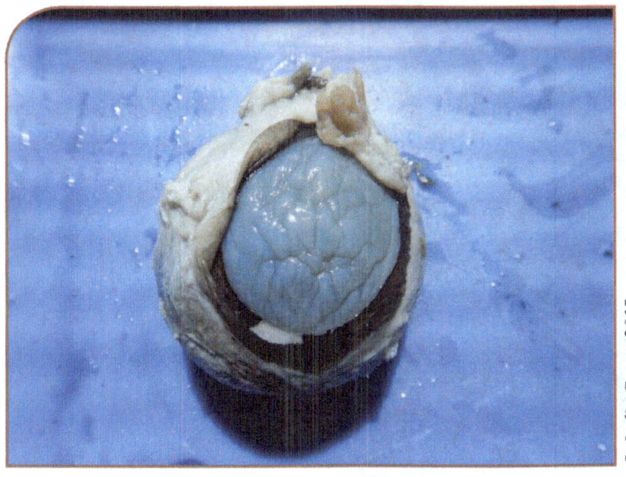

© Jodie Gerts, 2015

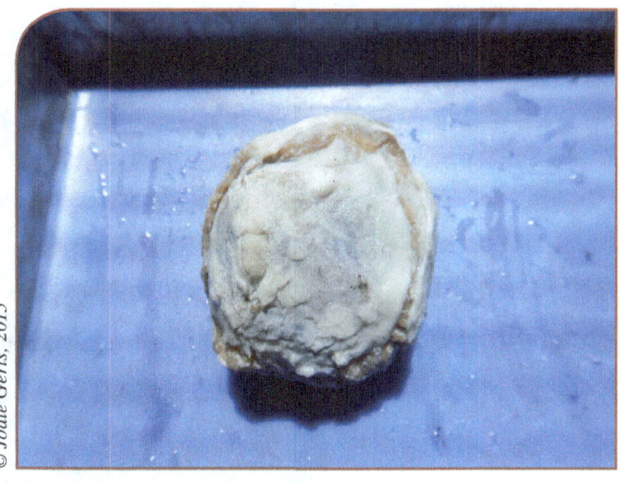

© Jodie Gerts, 2015

3. Remove the **EOM** and **orbital fat** using scissors being careful to isolate the **optic nerve**. This will expose the thick, white sclera. Do not puncture eye.

4. Cut a slit in the periphery of the **cornea** between the cornea and sclera using a scalpel. Slit should be about 1 cm long. This releases the **aqueous humor**.

5. Cut a 1 cm long slit using the scalpel in the midline of the **sclera** halfway between the cornea and the optic nerve.

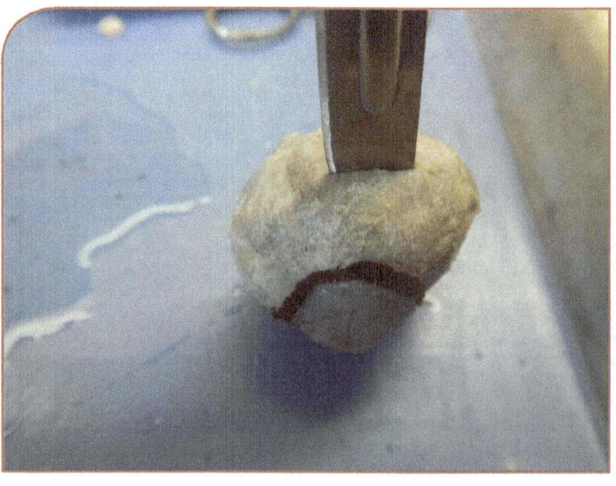

6. Inserting scissors into the slit, circumferentially cut around the sclera. The jelly-like **vitreous humor** attached to the lens will be released.

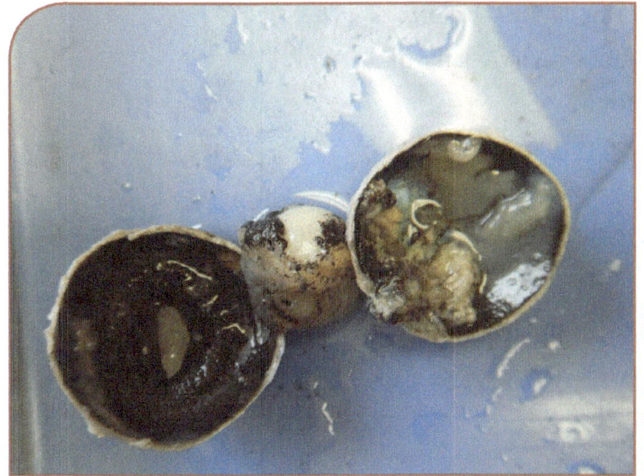

7. Insert the scissors into the previous corneal cut and remove the entire cornea by cutting circumferentially around the cornea.

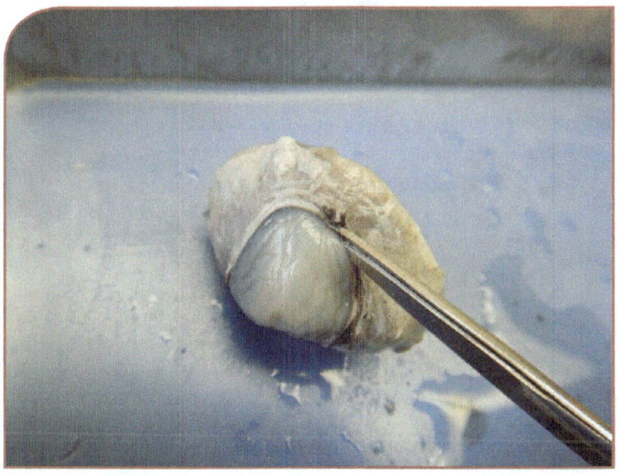

8. Pull off the **ciliary body** at the ora serrata using forceps.
9. Identify structures by pinning them with the numbers provided.
10. Check off with the instructor.
11. Clean your area using disinfectant and towels and dispose of the eye.

Pin these structures

1. EOM (red muscle) and orbital fat.
2. Aqueous humor in the anterior cavity—watery.
3. Vitreous humor in the posterior cavity—jelly-like.
4. Cornea—clear, tough covering.
5. Lens—hard, round part. Looks like a small grape.
6. Sclera—the white part of the eye.
7. Iris—black tissue of the eye that contains curved muscle fibers.
8. Retina—yellow, gooey part; tissue in the back of the eye where light is focused; connects to the optic nerve.
9. Optic nerve.
10. Optic disk.
11. Choroid—pigmented tissue deep to the retina (attached to the choroid is the tapetum lucidum—the opalized, reflective part).
12. Ciliary body—dark, ridged structure over iris, located on the back of the iris that has muscle fibers to change the shape of the lens.
13. Ora serrata—scalloped anterior margin of the retina.

PART 2:

EAR AND EYE MODELS

Ear Model

- A. External ear
- B. Middle ear
- C. Inner ear
- 1. Auricle
- 3. Tympanic membrane
- 7. Auditory tube or eustacian tube
- 8. Malleus
- 9. Incus
- 10. Stapes
- 12. Utricle
- 13. Oval window
- 14. Round window
- 15, 16, 17. Semicircular ducts
- 18. Cochlea
- 19. Vestibulocochear nerve

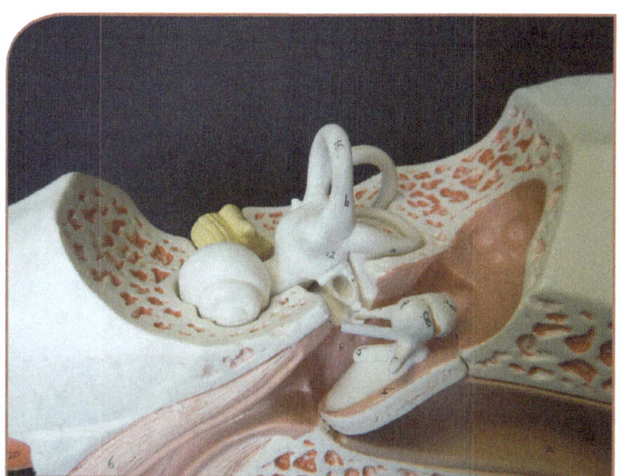

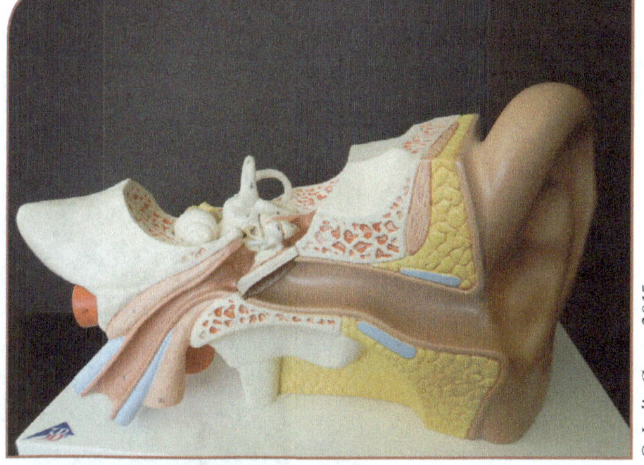

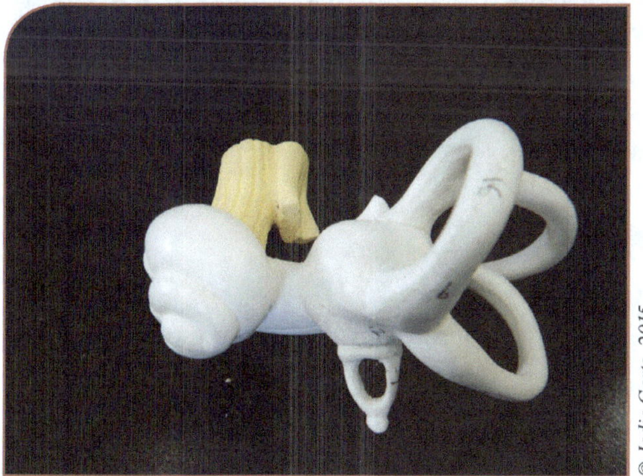

Eye Model

Anterior cavity

Choroid

Ciliary body

Cornea

EOM

Iris

Lacrimal gland

Lacrimal sac

Lens

Optic nerve

Ora serrata

Posterior cavity

Pupil

Sclera

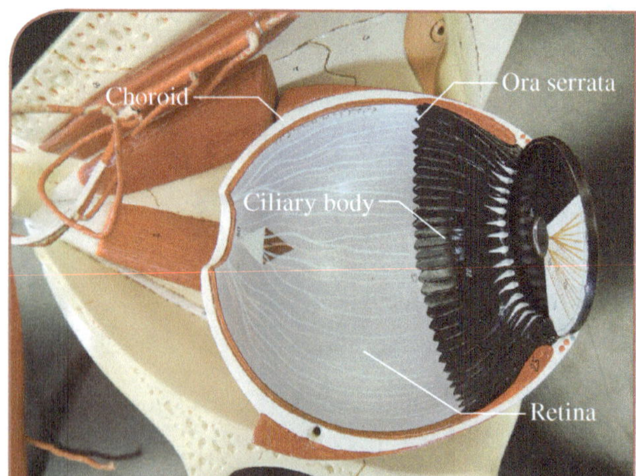

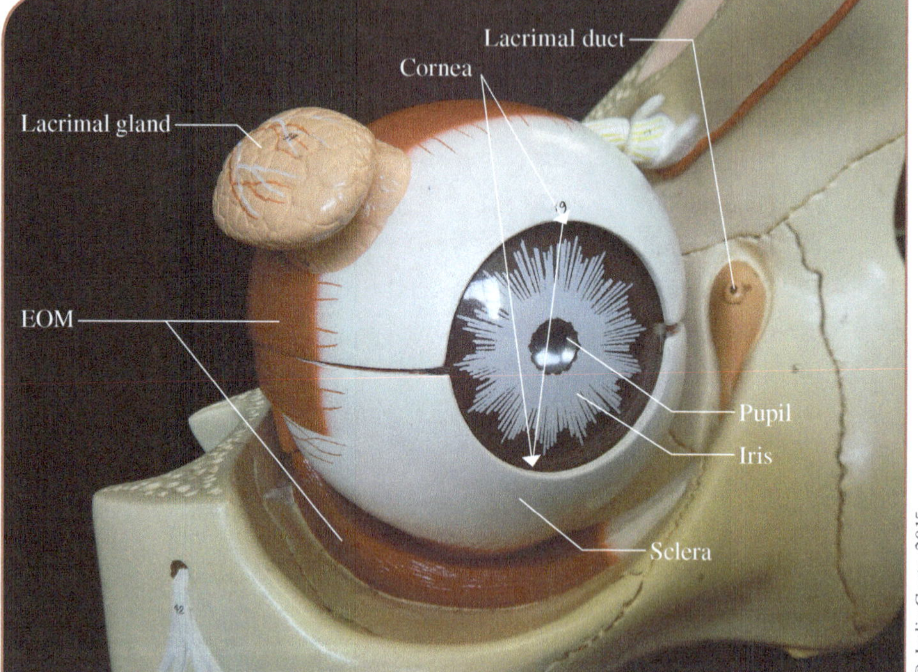

PART 3:

SEMMES-WEINSTEIN MONOFILAMENT TESTING

The Semmes-Weinstein monofilament form of testing provides information about peripheral neuropathy and loss of protective sensation. We will test for general sensation of the foot using the monofilaments. According to the American Diabetes Association, all patients with diabetes should be screened for loss of protective sensation in their feet (peripheral neuropathy) when they are initially diagnosed and then annually thereafter.

The Semmes-Weinstein 5.07 monofilament nylon wire exerts 10 g of force when bowed into a C shape against the skin for 1 second. Patients who cannot reliably detect application of the 5.07, 10 g monofilament to designated sites on the plantar surface of their feet are considered to have lost protective sensation. Neuropathy usually starts in the toes and progresses to the metatarsal heads. It is likely that these areas will be the first to show neuropathy with the Semmes-Weinstein monofilament exam. There are different methods for implementing this test, we will follow the most efficient method using four test sites on the plantar foot.

Procedure

1. Place the patient in supine position with shoes and socks removed.

2. Tell the patient that you are testing for loss of protective sensation, which increases the risk of foot ulcers and amputation.

3. **Begin with the 5.07 (or 10 g) monofilament.** Touch the monofilament wire to the patient's skin on the arm or hand to demonstrate what the touch feels like.

4. Instruct the patient to respond "Yes" each time he or she feels the pressure of the monofilament on the foot during the exam.

5. Instruct the patient to close his or her eyes and keep toes pointing straight up during the exam.

6. Hold the monofilament perpendicular to the patient's foot. Press it against the foot, increasing the pressure until the monofilament bends into a C shape. (The patient should sense the monofilament by the time it bows.) Only allow the monofilament point to touch the foot or the test is invalid.

7. Hold the monofilament in place for about 1 second and remove.

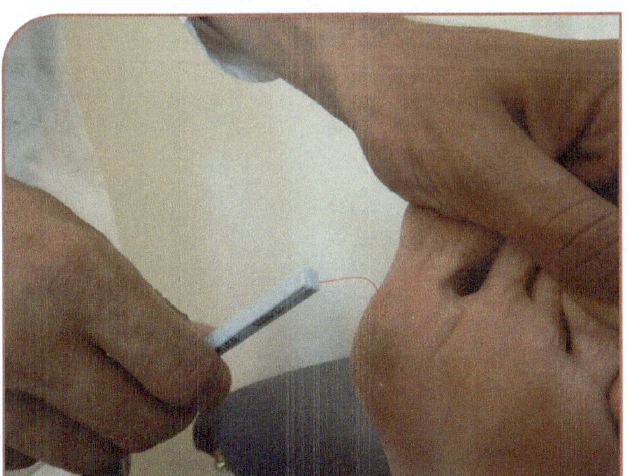

© Jodie Gerts, 2015

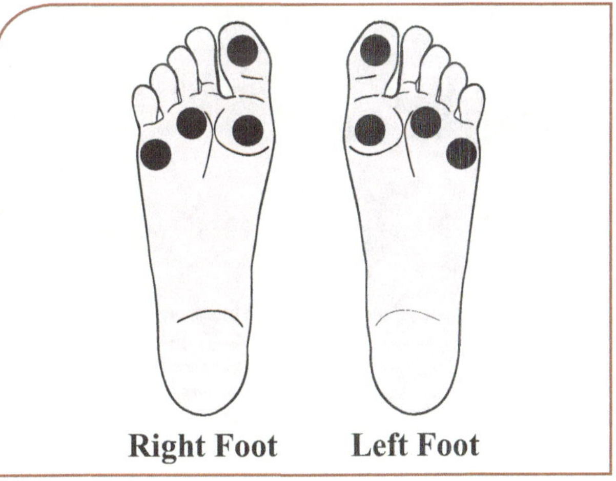

© laschi/Shutterstock.com

8. Locations for testing: On both feet, use the first, third, and fifth metatarsal heads and plantar surface of the distal hallux. Avoid callused areas.

9. Record response on foot screening form with "+" for yes and "−" for no.

10. If the patient does not feel the point, move to the next largest monofilament. Clinically if the patient can feel the 10 g force the test is terminated. However, in this class, move to the next smallest monofilament to continue the test to determine your patients level of sensation.

Location	300 g		10 g		4.0 g		2.0 g		0.4 g		0.07 g	
	Right	Left	Right	Left	Right	Left	Right	Left	Right	Left	Right	Left
1st metatarsal												
3rd metatarsal												
5th metatarsal												
Distal hallux												

Normative Age Values for Superficial Sensation

Age: 18–34: 0.4 g of force

Age: 35–64: 2.0 g of force

PART 4:

WORKSHEET ON SYMPATHETIC NERVOUS SYSTEM

Draw in the sympathetic nervous system template the spinal route of the preganglionic and postganglionic fibers to the skin arrector pili muscles, sweat glands and blood vessels of the extremities. Identify the approximate spinal cord level where the preganglionic fiber cell body originates. Identify the white and grey rami communicans.

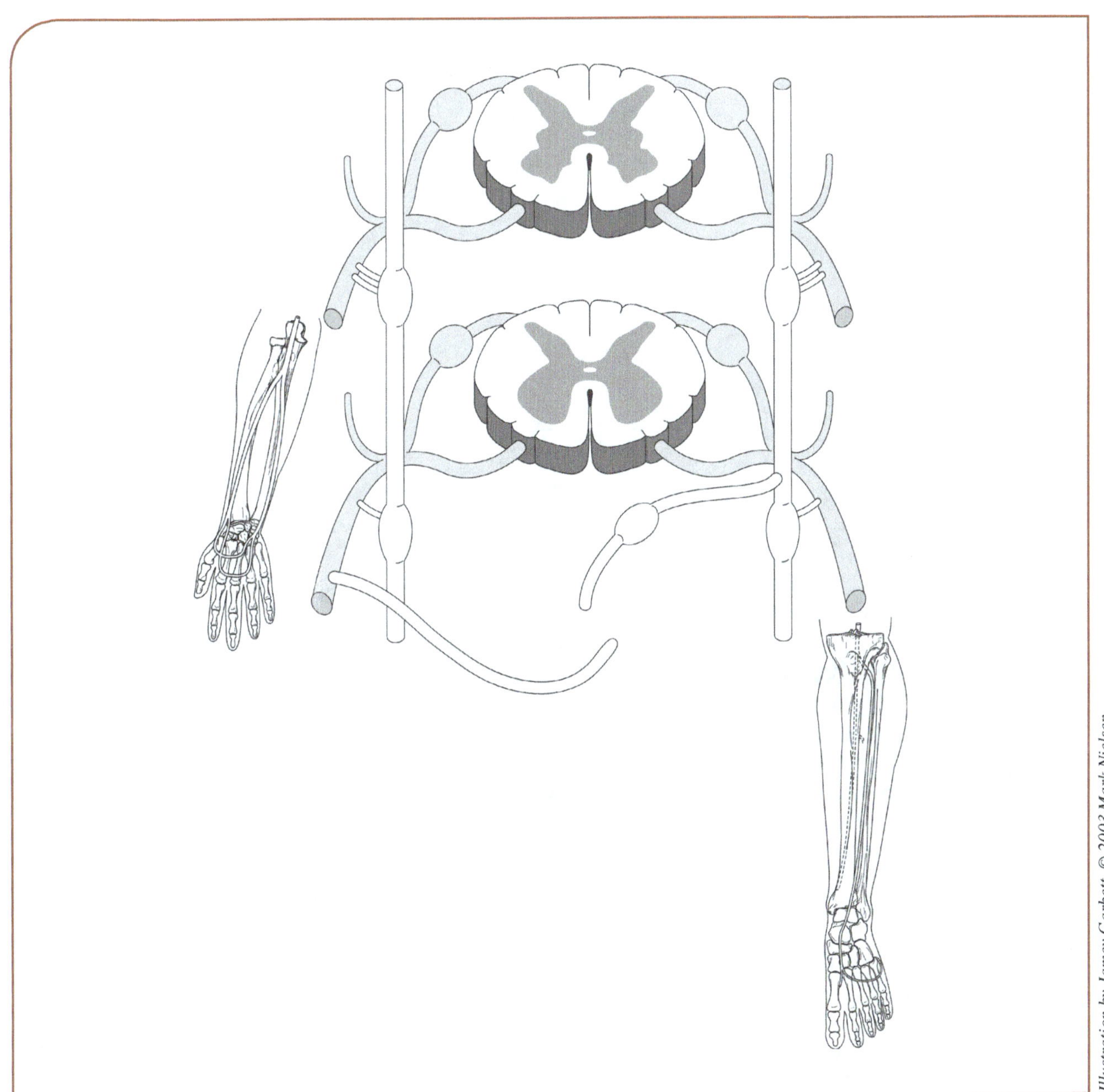